Biodiversity

Edmundas Lekevičius

Biodiversity

Maintenance, Function, Origin, and Self-Organisation into Life-Support Systems

Springer

Edmundas Lekevičius
Nature Research Centre
Vilnius, Lithuania

ISBN 978-3-031-11581-3 ISBN 978-3-031-11582-0 (eBook)
https://doi.org/10.1007/978-3-031-11582-0

This Springer imprint is published by the registered company Springer Nature Switzerland AG
The registered company address is: Gewerbestrasse 11, 6330 Cham, Switzerland

In memory of Edward Osborne Wilson, one of the greatest biologists of our time.

Foreword

Modern evolutionary biology has been tremendously effective at understanding and dissecting the mechanisms at work in the evolution of species, but this progress has largely come at the expense of explaining the broader patterns and processes of diversification of life on Earth. Indeed, the so-called modern synthesis of evolution focuses almost entirely on basic evolutionary mechanisms such as mutation, natural selection, drift, and gene flow, while it has little to say about broader questions such as: Why are there so many species on Earth? Why and how have ecosystems evolved to their present structure, functional organisation, and level of complexity? What is the functional role of genetic, phenotypic, and species diversity in the functioning of ecosystems and the biosphere as a whole? Yet these questions are clearly of the utmost importance in the current era of massive anthropogenic environmental changes, which some have called the 'Anthropocene'.

Based on this observation, Ed Wilson (1992), in his book *The Diversity of Life*, predicated that 'the greatest advances in evolutionary biology will be made in ecology, explaining more fully in time why the diversity of life is of such and such a nature and not some other'. In the present book, Edmundas Lekevicius takes up this challenge and proposes a synthetic view of the ecological patterns and processes that are involved in the evolution and diversification of life on Earth, from the small-scale interplay of competition and cooperation between species to the large-scale self-organisation of biodiversity and ecosystems through Earth's history. Although Lekevicius ironically refers to himself as a 'sceptical maverick' in modern evolutionary biology, I believe his book asks the right questions and is fully in line with Darwin, who did not hesitate to devote entire books to the emergence of sociality and morality in humans or to the role of earthworms in the economy of nature.

Almost inevitably, on a subject as vast as the maintenance, function, origin, and self-organisation of biodiversity, Lekevicius' book raises more questions than it answers. To take a few examples among many, his book argues convincingly that some sort of trophic or functional division of labour is necessary for an ecosystem to function, but it does not say roughly how many species or phenotypes would be needed to do so. Similarly, it argues convincingly that a diversity of species or phenotypes is necessary to cope with spatially and temporally varying environments, but it does not provide even a rough estimate of the number of species or phenotypes needed for this purpose. Ecology simply does not yet have the theoretical and empirical elements to answer such questions quantitatively. As a result of this lack of knowledge, it is still impossible to properly test many of the predictions derived from either theoretical deduction or empirical induction regarding the level of biodiversity on our planet, from the scale of a tiny microcosm to the scale of the entire biosphere. The subtitle of Lekevicius' book, 'Towards a general theory of biodiversity', aptly recognises the fact that we do not yet have all the elements of a truly general theory of biodiversity.

Lekevicius' philosophical and methodological stance, however, is sometimes problematic in this respect. For instance, he argues that coping with such a complex phenomenon as biological diversity requires conceptual modelling, not mathematical modelling. As a theoretical ecologist who has developed theory on many of the issues discussed in this book, I respectfully disagree with this view. On the contrary, my experience has shown me that formalising concepts into mathematical models is of paramount importance in clarifying concepts, turning them into testable predictions, and then testing these predictions either experimentally or by using an appropriate combination of observational data and clever statistical analyses. In doing so, scientific progress is greatly accelerated at relatively low cost (Loreau, 2010b).

Lekevicius is also highly critical of evolutionary biologists who have argued that the history of life is not necessarily progressive and that evolution has a large element of chance. In contrast to this view, he argues that life on Earth has a goal, to increase the total biomass of the biosphere, and that evolution is constantly moving life towards this goal. I have explained in detail elsewhere (Loreau, 2010a) why we should not expect ecosystems and the biosphere to have a single goal because of the tension between different levels of natural selection, and why we should carefully distinguish between large-scale trends in the development of ecosystems or life in general and true goals or optimum principles. The current disruptions brought about by our own species, which are reducing the total biomass of the biosphere, clearly illustrate that life does not always proceed gradually or progressively. More generally, I strongly believe that chance and necessity are only abstract concepts whose relevance depends heavily on the spatial and temporal scale considered, and thus that a complete theory of biodiversity requires an explicit consideration of the issue of scale and an explicit integration of processes across scales.

Despite these reservations, I thoroughly enjoyed reading Lekevicius' book. His book is pleasantly provocative, full of stimulating ideas and established ecological facts, and it shakes up neo-Darwinism in a refreshing way. It reminds us that some of the biggest questions that evolutionary biology should be answering remain unanswered, even though these questions are essential to overcome the current global biodiversity crisis.

Centre for Biodiversity Theory and Modelling in Moulis (France)
Moulis, France
04.03.2022

Michel Loreau

References

Loreau, M. (2010a). *From populations to ecosystems: Theoretical foundations for a new ecological synthesis* (Monographs in population biology). Princeton University Press.
Loreau, M. (2010b). Linking biodiversity and ecosystems: Towards a unifying ecological theory. *Philosophical Transactions of the Royal Society B, 365*, 49–60.
Wilson, E. O. (1992). *The diversity of life*. W. W. Norton.

Acknowledgments

This book would hardly have been completed without the support of my colleagues from the Nature Research Centre of Lithuania. My foremost thanks go to Vincas Būda, the director of this Centre, and Ričardas Paškauskas, the head of the Laboratory of Algology and Microbial Ecology at this Centre, who both have not only kept me going and would not allow me to sink into desperation but also managed to organise funding for the preparation of this book. I also highly appreciate the especially valuable mediation and advice provided by Gediminas Valkiūnas, the head of the P. B. Šivickis Laboratory of Parasitology at the Centre. Finally, I would like to express deep gratitude to my colleague Yuri Gleba from Nomad Bioscience GmbH in Germany for his recommendations and comments.

I am equally grateful to my translator Mantas Zurba – I was impressed by his dedication and diligence. He did more than just translate the book, as he helped me to correct numerous inconsistencies in the original text.

Edmundas Lekevičius

Contents

About the Author

Edmundas Lekevičius was born on 7 December 1942 in Alytus, Lithuania. In 1961, he enrolled at Vilnius University, majoring in biology. In 1962, he was drafted into the Soviet Army, from which he was discharged in 1965, and continued his studies at Vilnius University. In 1971, he began his postgraduate studies at the Institute of General Genetics, Moscow, and successfully defended his PhD dissertation in 1974. Until 1984, he worked as research fellow at the Institute of Botany, Vilnius. Later, until 1991, he headed the Laboratory of Hydrobiology at the Institute of Zoology and Parasitology, Vilnius.He published in 1986 the book *Elements of a General Adaptation Theory* (in Russian), which concretised and elaborated on Michael Conrad's Hierarchical Adaptability Theory developed in 1983. After Lithuania regained its independence, Lekevičius was appointed advisor to the Lithuanian Government. From 1992 until 1998, he led the reform of science education in Lithuanian general education schools. His postdoctoral (habilitation) work *A Model of Adaptive Rearrangements on Community, Population, and Organism Levels* was successfully defended in 1997. From 1998 to 2013, he was a professor at Vilnius University teaching *general ecology*, *adaptation theory*, and *ecosystem evolution*.In 2002, he published the book *The Origin of Ecosystems by Means of Natural Selection*, arguing that, in addition to the evolution of species, there was also the evolution of ecosystems that does not require the selection of ecosystems as units. Lekevičius considers his greatest contribution to biological science his discovery of a way of integrating ecology and evolutionary theory, thus solving the riddle of biodiversity.

Part I
In Search for a Functional Explanation

Chapter 1
Introduction: Contemporary Biology Struggles to Explain the Great Diversity of Genotypes and Species

Still a student, together with all my fellow biologists I attended lectures on evolutionary biology. I must admit that by now I have completely forgotten everything that our professor was hammering into our heads. All that remains is a very general impression that no branch of biology could be a more boring subject. Darwin provided a perfect explanation on how and why everything we see around us has evolved. A few minor gaps remaining in this understanding have eventually been filled by geneticists. Apparently, there was nothing left to do in this branch of biology but to memorise the key postulates. Just to pass the exam. Of course, back then, I was too young to know the simple wisdom: if all major issues appear to be 'clear' to all experts in a particular field, it may be time to ring the alarm bells because something must be definitely wrong.

I did not have to wait long for the proof of that wisdom. First, in about 1978, I learned that almost all studied animal and plant populations – and there were hundreds of them – featured a high diversity of genotypes (polymorphism), while most individuals within them turned out to be heterozygous. As if that were not enough, some studies demonstrated (Lewontin, 1974) that at least part of that diversity is supported by natural selection and passed on from generation to generation. But wait, we all 'knew for sure' that this could not be the case, that natural selection was supposed to eliminate from the population all genotypes but one, that with the highest adaptive value. Before the above-mentioned discovery, natural selection was considered the major clean-up engine, the antithesis of diversity rather than its supporting force. Darwin himself would surely have been surprised by the existence of polymorphism. Thus, we could witness the occurrence of a paradox currently known as the paradox of genetic diversity.

It was roughly then, that I became interested in hydrobiology where I found a description of a similar paradox of the plankton (Hutchinson, 1959, 1961). The author, a world-famous ecologist, was asking how come that a single drop of lake water contains a few dozens of algae and cyanobacteria species. It appeared 'obvious' that all these species needed the same resources: light and inorganic nutrients or biogenic substances. Consequently, they should occupy the same ecological

E. Lekevičius, *Biodiversity*, https://doi.org/10.1007/978-3-031-11582-0_1

niche. Why have the stronger species not outcompeted the weaker ones, as postulated by Gause's principle, well-known among ecologists?

Eventually, I have discovered the existence of yet another paradox, the paradox of the meadow, analogous to the paradox of the plankton. The very fact that dozens of different plant species grow side by side in a meadow has been known for ages. Darwin (1872) did not seem to be bothered by that (p. 88):

> ...I found that a piece of turf, three feet by four in size, which had been exposed for many years to exactly the same conditions, supported twenty species of plants, and these belonged to eighteen genera and to eight orders, which shows how much these plants differed from each other.

He meant, in modern terms, that the niches of these plants were different and therefore they could coexist without problems. Unfortunately, we, the modern biologists, fail to see it with the same clarity. Just as in case of the plankton, we would prefer to see only one, the most competitive species instead of those twenty. We, ecologists, are not supposed to equate morphological differences with the functional ones, so we cannot accept Darwin's argument without proof. In this context, another fact that a mere hectare of the tropical rainforest can host a few hundred tree species appears even more confusing to us. Daughter species may emerge separately from the parental population, but sooner or later their ranges overlap, and, it turns out, they can peacefully coexist for millennia. If the species of the same guild are always competing for resources, why then the strongest competitors fail to phase out the weaker ones? Of course, we can assume that they have coexisted for ages because their niches only appear identical while in reality they have certain mismatches. However, that would merely be an unproved assumption. Scientists, just as police investigators, are seeking for undeniable 'evidence'.

Finally, while studying *The Origin of Species* ever deeper, I realised the existence of yet another conspicuous fact hardly compatible with the theory of natural selection. Darwin seemed to be especially puzzled by it. According to the logic of the theory of natural selection, highly organised life forms should be better adapted to the living conditions on Earth than more primitive ones, therefore they should have advantage in the struggle for existence. However, it was obvious even to an average biologist of Darwinian times that, despite their simplicity, primitive life forms are no less adapted to their environment than the highly organised species – each of them to their own specific conditions. This would imply that, through millions of years, evolution has not brought any progress, and the adaptedness of species has not been increasing throughout generations. Admittedly, our own species should be considered a special case possessing all the qualities of a super-competitor, but this could be interpreted as an exception to the general rule. Indeed, a supporter of the classical theory would hardly be able to apply the same statement to, say, a mould, a plant, or even an ape. Their competitive advantages over, say, a bacterium do not appear that obvious. Again, we encounter a sort of paradox: a respectful theory that should not be doubted predicts a certain outcome, while in reality we observe something different. Or at least something more than the theory predicts.

After long and painstaking contemplation, Darwin finally produced a rather peculiar explanation to the fact that some primitive species had survived from the very dawn of evolution without getting eliminated (Darwin, 1872, pp. 98, 100):

> [natural selection] 'only takes advantage of such variations as arise and are beneficial to each creature under its complex relations of life ‹...› But the main cause lies in the fact that under very simple conditions of life a high organisation would be of no service ‹...› If it were no advantage, these forms would be left, by natural selection, unimproved or but little improved, and might remain for indefinite ages in their present lowly condition'.

In other words, the existence of primitive organisms in contemporary nature should be explained by their primitive needs and equally unsophisticated activities. I am not sure about my reader, but I find the explanation unconvincing. Darwin himself appeared unconvinced, as he immediately attempted to contradict his own words: if 'a high organisation would be of no service' then why did it appear useful at the dawn of life which back then consisted exclusively of primitive forms? It must have been useful, as there is no other explanation to why primitive life forms evolved into more sophisticated ones. Admitting the limited possibilities of science in his era, Darwin concluded (Ibid., p. 100): '...as we have no facts to guide us, speculation on the subject is almost useless'.

In this monograph, I intend to discuss each of these paradoxes and attempt to demonstrate that they can be resolved or at least rendered less mysterious by adjusting some traditional concepts of biology. Let me list the key questions addressed in the subsequent chapters:

- How has a diversity of species occurred and increased, and why has life not taken an opposite course leading to a sort of superorganism comparable to the intelligent ocean of the science fiction novel *Solaris* by S. Lem?
- What forces maintain the diversity of genotypes and species?
- What is the role of that diversity in the daily functioning of life?
- Why has species diversity acquired its present functional pattern, and not some alternative ones (in terms of standard nutrient cycles and ecological pyramids)?

The reader may be puzzled somewhat by the first topic which does not seem to have been discussed in this context in any sources. When we, theoretical biologists, ask why diversity is needed in nature, or why so many species are found in our natural environment, it is not just about whether the mechanism that produces two species out of one is sympatric or allopatric; not just about mutations and recombinations leading towards a new species and occurrence of reproductive isolation. The mystery of biological diversity would have been resolved long ago if it could be reduced to such a limited set of questions. First, a deep-thinking theoretician cannot take for granted the path of splitting into species and individuals, chosen by the life on Earth since its very occurrence, instead of spreading across the water column or the surface of land as a uniform organic film. The latter scenario would have produced neither species nor multiple individuals, but rather one or several superorganisms akin to that intelligent ocean in Lem's *Solaris*. Such superorganisms would have probably been very efficient as they would not have wasted energy and materials on continuous struggle and competition. Of course, this and similar questions of the 'what if …?' category can only be answered by using a deductive method highly favoured by the scholars of hard sciences.

These four topics could be assigned to the field already described by some authors (Cardinale, 2013) as the general theory of biodiversity. It would be premature to claim that the theory is already accomplished, but its outlines are becoming clearly visible (Lekevičius, 2018a, b), and it is them that I intend to present and explicate to my readers in this monograph.

I feel I should try to explain here why this theory is being produced only now, some 160 years after the publication of *The Origin of Species*, despite its indisputable importance to life sciences. I see no easy answer, since I am not an expert in the history of science; moreover, there may be at least several different explanations. To start with, one should not get the impression that some crucial empirical evidence necessary for the integrative effort of such a grand scale has surfaced only in the last few years. I am deeply convinced that this work could have commenced some 50–60 years ago. Maybe such a synthesis was not needed then? I think the answer will only arrive together with the final product. Therefore, this argument does not seem appropriate either. Much more convincing is the explanation that the phenomenon of biodiversity appears too complex to many of us, therefore we prefer to analyse it part by part, and this is exactly what is happening with various degrees of success. In my opinion, however, perception of complexity is quite subjective, and a lot depends on the chosen method of study. If certain issues appear too complex to some researchers, they may have chosen a wrong approach, wearing 'a wrong pair of glasses' so to say…

I prefer the following explanation. None of us is capable of completely and adequately assessing the totality of things that we do not know. The absence of questions of principal importance in a certain scientific field does not always mean that all the problems have been resolved. Now and then this may indicate something else, namely, a crisis of certain extent, intellectual stagnation. Historians of science are well familiar with such situations. For example, merely a few centuries ago it was 'a common knowledge' that the Sun was revolving around the Earth, and not the other way around as suggested by some 'weirdos'. Nothing can suppress thinking more than the trivial 'self-evident facts'. However, any 'fact' becomes a scientific fact only when it is linked with other facts or, in other words, is given an explanation.

In my opinion, this is exactly what happened with the problem of biological diversity. Darwin might have failed to find the answers he would deem satisfactory, but we might have forgotten the very questions, thus failing to match his broadness of thinking. When no more questions were posed, the problem itself disappeared from publications. Only a handful of restless minds seem to persist in stirring up trouble. Even this explanation, however, appears too general and rather lame, as it itself calls for clarification.

I must admit I am rather sceptical about the moods and approaches prevailing in modern biology. For instance, I tend to disagree with the opinion of many evolutionary biologists (or, perhaps, a vast majority of them) that evolutionary biology has already provided answers to its essential questions, and all that remains is clarification of various details. For example, Kutschera and Niklas (2004) see the near future and the progress of this discipline mostly in the successful development of mathematical simulations (digital organisms), molecular biology (epigenetics, gene duplications), development biology (Hox-genes), paleobiology (rates of evolution), and ethology (sexual selection). Quite symptomatic is the fact that these and five other fields, suggested by these authors as the most promising, include neither ecology of ecosystems, nor any reference to the conundrum of biodiversity and the related paradoxes! A number of other recognised experts seem to be sharing this or similar opinion, among them Dawkins (1976), Maynard Smith and Szathmáry (1995), Ridley (2004), Pigliucci (2009), Futuyma (2009, 2015), Freeman and Herron (2014), and Laland et al. (2015). (I am not claiming hereby that all these authors share an identical position regarding the achievements and perspectives of evolutionary biology.)

It seems that modern evolutionary biologists are not preoccupied by the questions *how* the surrounding world with its entire diversity has come into being and especially *why* it has acquired its present form, rather than a different one. I am almost sure that, should some maverick pose these questions at some conference, he or she would provoke broad bewilderment, perhaps even scornful remarks, and eventually would receive a response along these lines: 'Dear colleague, you are breaking in through an open door; your questions have already been answered, and you should familiarise yourself with publications that are long considered classical works'. A less arrogant listener might comment instead: 'Dear colleague, it is quite obvious that diversification is caused by disruptive selection, this was proven long ago; as for your question about the very existence of diversity, I find it simply incorrect; this is a question I would rather expect from a philosopher'.

Nevertheless, I seem to belong to a broader circle of sceptical 'mavericks', and I am extremely pleased to find one of the most outstanding modern biologists, Wilson (1992), among them. He wrote the following (p. 85):

> What we understand best about evolution is mostly genetic, and what we understand least is mostly ecological. I will go further and suggest that the major remaining questions of evolutionary biology are ecological rather than genetic in content. They have to do with selection pressures from the environment as revealed by the histories of particular lineages, not with genetic mechanisms of the most general nature (…) I think the greatest advances in evolutionary biology will be made in ecology, explaining more fully in time why the diversity of life is of such and such a nature and not some other.

The fact that neo-Darwinism ignores some questions which are by no means of secondary importance had been highlighted even earlier by Waddington (1969), a famous organiser of the movement for the new type of theoretical biology. He called for the creation of a 'post-neo-Darwinism paradigm', i.e. of a methodology that would not allow to ignore the species interactions within communities and ecosystems. It is these interactions that determine the evolutionary vectors of individual species.

And yet, should anybody ask me whether on top of that I consider myself anti-Darwinist, as many of my colleagues are convinced, my answer would be negative.

These accusations remind me of some facts of my biography. To avoid any confusion, I open one of the issues of *Mokslas ir gyvenimas* (*Science and Life*) magazine (1984) and find my own article under the title *Ko nepasakė Darvinas* (*The things Darwin never said*). I wrote there that I had not been able to find any explanation of the astonishing diversity of the natural world comprising millions of species and billions of different ecotypes, genotypes, biotypes, etc. In order not to be labelled a grumpy critic, I attempted to provide my own answers in the same article: perhaps this could be explained by admitting that, in addition to the 'struggle for existence' profoundly emphasised by Darwin, there is also cooperation in nature. While plants may appear not so cooperative, the same cannot be said about animals. By no means, I claimed in this article that Darwin had been unaware of cooperation, but I was absolutely convinced that there was some discrepancy between the broadly accepted theory of natural selection and the real facts. Logical reasoning should suggest that overarching 'struggle' or competition should lead to the decrease of biodiversity, in other words, to the survival of a limited number of the best-adapted genotypes or species. But how are we supposed to explain biodiversity which seems to have been steadily increasing over those few billion years despite some mass extinctions?

I remember that this article provoked some harsh reactions from several professors who were then working at Vilnius University and who certainly considered themselves loyal Darwinists. Lekevičius was alleged to be a pure anti-Darwinist striving to undermine the very fundaments of modern biology. There were even some attempts to punish me in various ways, to impose administrative sanctions via some intermediaries, since I was working in another institution beyond their control. Bearing in mind the prevailing ideology of that time, I could not be – and was not – surprised by such a disproportionate reaction. I find this episode worth mentioning only insofar as whether my opponents correctly tagged me as anti-Darwinist. Now I know for sure: I might have been slightly too critical towards Darwin back then, but nowadays I consider myself not only a positive critic but also a successor and even defender of his ideas. By the way, they have to be defended not only from casual opponents but also from some acknowledged global-level experts setting trends in palaeontology and evolutionary biology (more on that in Chap. 21).

It is true that Darwin posed many more questions of fundamental importance than he could provide soundly based answers to. As I have already mentioned, the questions related to biodiversity appeared especially challenging to him. One may often hear that he only wrote about species improving from generation to generation thanks to ever-better adaptation to their environment. He allegedly ignored the diversification of species and their increasing abundance. And though his major work bears the title *The Origin of Species*, that title is misleading. Even the above-mentioned Wilson (1992) supports this opinion. I do not share it, however. A diligent examination of Darwin's (1872) work reveals some modest references to the increasing number of species. For instance, he writes (p. 103) that natural selection not only discards less adapted 'varieties' but also has another role to play:

> Natural selection, also, leads to divergence of character; for the more organic beings diverge in structure, habits, and constitution, by so much the more can a large number be supported on the same area (…) Therefore, during the modification of the descendants of any one species, and during the incessant struggle of all species to increase in numbers, the more diversified the descendants become, the better will be their chance of success in the battle

> of life. Thus, the small differences distinguishing varieties of the same species, steadily tend to increase, till they equal the greater differences between species of the same genus, or even of distinct genera.

The diverged 'varieties' and species occupy different positions 'in the polity of nature' (i.e. in the population and ecosystem), thus ceasing to compete among themselves and switching to peaceful coexistence. In other words, those less adapted can avoid elimination by engaging in different activity which better matches their needs, and wherein they have an opportunity to gain advantage over their competitors. Just as in economics: if you lose competition in one area, find another niche where you may become the best. Apparently, this principle applies not only to single individuals but also to the competition between different species. Consequently, Darwin may be considered to have pawed the ways leading to the explanation of intra- and interspecific specialisation, also known as niche segregation or niche partitioning. He even produced a conceptual model of species diversification which is presented in a form of chart and is the only illustration in the book (pp. 90–91).

Modern evolutionary biologists use another term, *character displacement*, instead of *divergence of character* but the main idea remains unchanged: they believe that divergence of certain features is caused by disruptive selection which eliminates intermediary 'varieties'. Nonetheless, as Pfennig and Pfennig (2010) conclude,

> Darwin's intuition was essentially correct: selection acting to minimise costly interactions between organisms can indeed promote divergent evolution and speciation. Yet (…) Darwin failed to appreciate the importance of divergence in reproductive characters, the effects of character displacement on sexual selection, and the importance of understanding the proximate mechanisms by which character displacement occurs. Nevertheless, Darwin's insights into the role of competition in divergence were well founded.

I also support this idea. Therefore, together with these and other evolutionary biologists (e.g. Boero, 2015), I tend to emphasise Darwin's unprecedented (even among modern biologists) insightfulness, rather than to downplay him. As for those who are so keen to attach tags and labels, I suggest classifying me, if necessary, as a rebellious, non-conjunctural or any other type of theorist, but by no means an anti-Darwinist – I find this label unfair and offensive. Yes, Darwin failed to provide answers to certain questions, and it is our duty to build upon his work. The paradoxes I mentioned before provide a good basis for future efforts. With his devotion to finding the truth, Darwin himself would encourage us to take this task on. This is exactly what I have been trying to do within the last 40 years. On the other hand, a reader of this monograph will find practically no ideas that have not been approved in scientific publications, no matter how controversial they may seem to some people.

References

Boero, F. (2015). From Darwin's Origin of Species toward a theory of natural history. *F1000Prime Reports 2015, 7*, 49. https://doi.org/10.12703/P7-49

Cardinale, B. J. (2013). Towards a general theory of biodiversity for the Anthropocene. *Elementa: Science of the Anthropocene, 1*, 000014.

Darwin, C. (1872). *The origin of species by means of natural selection* (6th ed.). John Murray. (Printed in 1998 by Cox & Wyman. Reading, England).
Dawkins, R. (1976). *The selfish gene*. Oxford University Press.
Freeman, S., & Herron, J. C. (2014). *Evolutionary analysis* (5th ed.). Pearson.
Futuyma, D. J. (2009). *Evolution* (2nd ed.). Sinauer Associates, Inc.
Futuyma, D. J. (2015). Can modern evolutionary theory explain macroevolution. In E. Serrelli & N. Gontier (Eds.), *Macroevolution* (pp. 29–85). Springer.
Hutchinson, G. E. (1959). Homage to Santa Rosalia or why are there so many kinds of animals? *The American Naturalist, 93*, 145–159.
Hutchinson, G. E. (1961). The paradox of the plankton. *The American Naturalist, 95*, 137–145.
Kutschera, U., & Niklas, K. J. (2004). The modern theory of biological evolution: An expanded synthesis. *Die Naturwissenschaften, 91*, 255–276.
Laland, K. N., Uller, T., Feldman, M. W., Sterelny, K., Müller, G. B., Moczek, A., Jablonka, E., & Odling-Smee, J. (2015). The extended evolutionary synthesis: Its structure, assumptions and predictions. *Proceedings of the Royal Society of London. Series B, 282*, 20151019.
Lekevičius, E. (Лекявичюс Э.) (2018a). Biodiversity – What for and why? I. In search for a functional explanation. *Journal of General Biology (Moscow), 79*, 201–220 (in Russian).
Lekevičius, E. (Лекявичюс Э.) (2018b). Biodiversity – What for and why? II. In search for a synthetic explanation. *Journal of General Biology (Moscow), 79*, 221–236 (in Russian).
Lewontin, R. C. (1974). *The genetic basis of evolutionary change*. Columbia University Press.
Maynard, S. J., & Szathmάry, E. (1995). *The major transitions in evolution*. W. H. Freeman.
Pfennig, D. W., & Pfennig, K. S. (2010). Character displacement and the origins of diversity. *The American Naturalist, 176*(Suppl 1), S26–S44.
Pigliucci, M. (2009). An extended synthesis for evolutionary biology. *Annals of the New York Academy of Sciences, 1168*, 218–228.
Ridley, M. (2004). *Evolution* (3th ed.). Blackwell Publ.
Waddington, C. H. (1969). Paradigm for an evolutionary process. In C. H. Waddington (Ed.), *Towards a theoretical biology. Vol. 2. Sketches* (pp. 106–123). Edinburgh University Press.
Wilson, E. O. (1992). *The diversity of life*. Penguin Books.

Chapter 2
Methodological Excursion: An Apology for Theoretical Synthesis, Simplicity, Deductive Method, and Conceptual Modelling

Life sciences can be divided into two relatively independent branches: functional biology and developmental biology (Mayr, 1968). The first comprises such disciplines as biochemistry, physiology, and ecology, while the second analyses, among other processes, ontogeny, ecological succession, and includes evolutionary biology. Genetics belongs to and links both branches. Experts in functional biology often settle for functional explanations, while developmental biologists look for causality (Lekevičius, 1985; Looijen, 1998). Let us take a closer look into this. Usually, in order to explain a process, one is expected to discover its causes. A cause is deemed to be a process or a condition that leads to another process called the effect. The cause must always precede the effect in time.

Contrary to physics and chemistry, biology allows for functional explanations in addition to causal ones. They do not refer to the cause, the past events, but rather to spatial interactions, the function, in other words, the role played by a structure within the system (Fig. 2.1). For example, the questions 'Why a hand?' or 'How to explain the existence of a hand?' could be answered by describing the complete growth of a hand in the process of embryonic development, and that would be a causal explanation. However, we could choose the other way: to list the functions performed by a hand, its purpose, while completely ignoring the process of its occurrence. Moreover, such explanation seems not only quite normal to many of us but also sufficient. Similar trends can be observed in ecology: when we attempt to explain the presence of a species in a particular location, we usually refer to the habitat that meets the needs of that species and its position in the flows of energy and/or materials. When we analyse the division of labour in the population according to sex, caste, rank, etc., we also make assumptions about the role of such division by concluding that diversity within population ensures better reproduction, defence against adversaries, more efficient foraging or housing efforts, etc. Physicists and chemists do not use functional explanations because there is no true division of labour or subordination of functions and their hierarchy in non-living nature. Admittedly, such division is common in man-made products, and we are quite used to the explanation of the 'raison d'être' (why are they here?) of a machine, a gadget or a tool, as well as of their composite parts, thus referring to their functions rather than the process of their production. Such explanations are also common in our daily life. For instance, one may answer the question why the church bells are ringing by giving the direct cause: somebody is pulling on the ropes; however, the asking person must have expected a functional explanation.

E. Lekevičius, *Biodiversity*, https://doi.org/10.1007/978-3-031-11582-0_2

In order to provide the most comprehensive functional explanation, people sometimes refer not only to the role of a structure within a larger system but also to the functions of the constituent parts of that structure. For instance, in addition to the functions of a hand as a body part, they would describe the role and interactions of bones, muscles, and nerves within the hand while it performs certain activities. In this case we speak about inner mechanisms, detailed elaboration on the things that were initially presented only in general terms (Fig. 2.1). Such approach may lead us down to the role of individual cells or even macromolecules. Methodologists define this process as analysis. The opposite process wherein we use the data about constituent parts to understand the functioning of the entire system is called synthesis (Fig. 2.2).

The concepts of analysis and synthesis are linked with a couple of other terms, such as holistic, or systemic, and reductionist or, as I would prefer to call it, 'partistic' approach. The holists are seeking a generalised picture not overloaded with details; they sympathise with the ideas of synthesis of knowledge. A researcher striving only for a holistic description of life does not necessarily need to go into details of physiological and biochemical mechanisms. He or she then obtains a simplistic picture of life stripped of the zillions of details that characterise lower levels of organisation. Of course, such researchers remain free to 'dive deeper' and enhance their description with additional details. The ecosystem concept, for

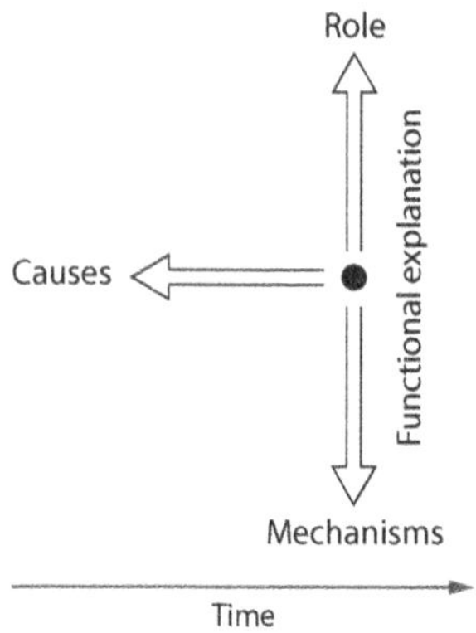

Fig. 2.1 Biologists and ecologists seek both causal and functional explanations. Sometimes a single explanation successfully encompasses both types, which is extremely useful for any integration attempt

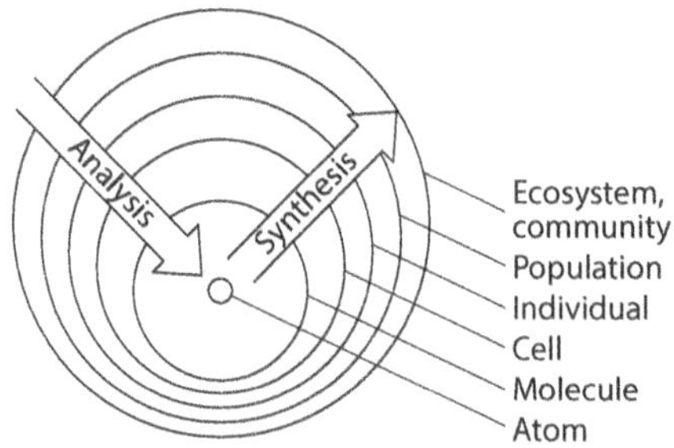

Fig. 2.2 Analysis allows to understand the parts of a system, while synthesis is necessary to get a holistic picture by using the data about the constituent parts

example, can be considered holistic, because it ignores even population-level processes, not to mention physiological or biochemical ones. This makes it so easy to comprehend and to be presented as a simple conceptual model. The core of Darwin's theory is also relatively simple because it is dissociated from physiology and the mechanisms of heredity and variability. Geneticists of the twentieth century who had 'submerged' down to the cell, chromosome, and gene levels mostly added more details to Darwinism, rather than expanded it. Whenever a new, broader version of Darwinism will occur (which I consider inevitable), explanations provided by geneticists will remain just as good and, most likely, they will play the same role of enhancing it with details. This is the meaning I attach to the principle of 'near-decomposability of systems' (according to H. A. Simon) well known to systemologists. On the other hand, the above-mentioned procedures would not be possible if another, even more fundamental principle did not apply to life – the principle of functional hierarchy (see also the next chapter).

Ecologists supporting 'partistic', or reductionist, views usually have little sentiment for a holistic picture as they consider it metaphysical, unpractical, or unattainable ('one cannot have it all'). Nowadays, this 'partistic' approach is especially popular both in ecology and in the theory of evolution. Nonetheless, in my monograph, I intend to present solutions that could be described as synthetic or holistic, and I shall do it on purpose with the aim to present the phenomenon of biodiversity in the simplest possible form. Simplicity does not mean lower quality. I shall attempt to persuade my reader that this phenomenon is not as complex as it may seem. Moreover, I am convinced that 'partistic' approach alone would never provide us with the answers to the questions posed in this monograph.

As I wrote in the introductory chapter, we are keen to discover what supports biodiversity and what is its functional role. Thus, we are seeking a functional explanation of this phenomenon. To achieve that, we shall have to take a closer look at functional biology. On the other hand, we have an ambition to shed some light also on the occurrence of this diversity, and this requires a causal description which can only be found in the developmental biology.

Eventually, we shall not be able to avoid the question about the link between the function and the cause, biological space, and biological time. Maybe functional biology is indeed detached from evolutionary biology and, despite obvious links between the two, one cannot be considered a derivative of the other, as many tend to think? I presented my own answer long ago, and I have not changed my opinion since: functional principles not only may, they must be incorporated into the explanans (explanatory statements) of evolutionary explanations; there is no other way. Only by choosing this path we shall be able to produce a conditionally complete (i.e. synthetic) explanation and simultaneously integrate the theory of evolution with functional biology (Lekevičius, 1985, 1986; see also Vermeij, 2004). It should be added that, in addition to the inductive method, e.g. fossils, we also have to use the deductive method to be able to describe evolution. I refer, first of all, to the principle of uniformitarianism or actualism. It postulates that the present is the key to the past: nature and its organisms might be changing, but not the laws of nature. The authors of this principle are geologists J. Hutton (1726–1797) and C. Layell (1797–1875), but Darwin is also known to have used it.

Darwin might be considered a pioneer in applying the ecological knowledge of his period to explain evolution. Let me briefly remind why. Upon his return to England in 1836, Darwin was almost convinced that species were not created by God as unchangeable entities. He had some supporting evidence, such as the similarities and differences of the Galapagos finches, as well as fossils collected by other researchers and Darwin himself. They all pointed to the idea that modern organisms could have evolved from more primitive ancestors, some of them extinct by then. But how was this transformation possible? Darwin had no idea where to look for the answers because, for obvious reasons, he could not observe evolutionary processes directly. He must have seen the only option and chose to use it: Darwin tried to find an answer in the present instead of digging into the past. He understood a lot by analysing the experience accumulated by selective breeders in producing new varieties of plants and animals. He concluded that various characteristics of living organisms could be modified by way of artificial selection. Of course, selection can only be successful if the chosen species comprises individuals with diverse characteristics, and the differences are at least partly inheritable. But what is the engine of selection in nature? In his quest for the answer, Darwin stumbled upon a book by famous economist and sociologist T. Malthus about the growth of human population leading to fight over resources. Darwin had the insight that not only people but also plants and animals 'struggle' against unfavourable environmental conditions but also for the living space and other resources. Thus, the concept of struggle for existence was created. Apparently, selection is driven by the same factors that limit the abundance of every species: shortage of food, competitors, predators, parasites, and adverse climatic conditions. Only the best fighters for existence survive, i.e. those best adapted to their environment. So, the idea of natural selection was inspired by the insight into analogy with struggle over resources and artificial selection. Analysis of Darwin's notebooks revealed (Schweber, 1977; Carey, 1998), that, while developing his ideas, he had also used the conclusions of another famous economist A. Smith, that unrestricted competition in economic life does not necessarily lead to the elimination of all but the strongest and most aggressive actors. Rather to the contrary, it leads to the division of labour, ever-growing diversity of specialised experts and economic progress. Just like in nature, concluded Darwin. See Lekevičius (2009) for more details on that.

It seems, Darwin has not derived his theory only by way of induction based on the data collected during his voyage aboard the 'Beagle', as it is often thought. He conceptualised it first, and only after having produced a hypothesis he tested the conclusions derived from it by way of deduction, using rich factual material. It is also obvious that his explananses of distant evolutionary phenomena always included statements regarding variability, heritability, and struggle for existence, i.e. the facts that seem to be taken from the present. Thus, he arrived at an integrated causal explanation of why species cannot remain unchanged through generations and why they evolve in a particular way. Could he somehow support his logical procedure by facts? Absolutely not. Just as I would not be able to do it, should I decide to make a statement that the driving force of evolution was cooperation rather than 'struggle'. Unfortunately, there seems to be no other, better-justified method. Interactions between organisms can hardly be assessed from fossils.

I am convinced that Darwin did not have a chance to resolve the problem of biological diversity despite his best intentions and efforts, because back then the concepts of ecological communities and ecosystems had yet to be developed. A rather primitive understanding of populations and their internal mechanisms was all that existed at that time. Therefore we, the successors of his ideas, must enhance his explanation of biodiversity with the modern interpretation offered by contemporary ecology. The latter first of all includes the concepts of communities and ecosystems. In conclusion, 'all we have to do' is create a new type of Darwinism, the 'eco-Darwinism'. This road is quite likely to finally lead us towards the answers to the question what produces diversity of species and genotypes and why it takes that and not some other form.

Has this Darwin's intuitive approach to the development of the theory of evolution ever been repeated in over 160 years? And, more importantly, has anybody attempted to integrate cooperation as the driving force of evolution acting instead of, or along with, 'struggle' into the explanans of the theory of evolution or its premises?

I am not sure about my reader, but I know just a mere handful of such attempts. One of them is the book by a Russian polymath and anarchist politician Kropotkin (1902) 'Mutual Aid: A Factor of Evolution'. Nowadays it has only historical value, therefore I am not going to analyse it. Another attempt is a truly revolutionary article published by Walker (1980) which I am going to discuss later in Chap. 18. Margulis and Sagan (1986) also proposed including the concept of positive interactions, coexistence, and cooperation into the theory of evolution in addition to competition and 'struggle for existence' (pp. 28–29):

> ... the view of evolution as chronic bloody competition among individuals and species, a popular distortion of Darwin's notion of "survival of the fittest", dissolves before a new view of continual cooperation, strong interaction, and mutual dependence among life forms. Life did not take over the globe by combat, but by networking. Life forms multiplied and complexified by co-opting others, not just by killing them.

This interpretation of evolutionary mechanisms is currently known as the theory of symbiogenesis (e.g. Corning, 2005; Guerrero et al., 2013). It can be briefly characterised as follows (Guerrero et al., 2013, p. 133):

> Symbiogenesis is the result of the permanent coexistence of various bionts to form the holobiont (namely, the host and its microbiota). The holobiome is the sum total of the component genomes in a eukaryotic organism; it comprises the genome of an individual member of a given taxon (the host genome) and the microbiome (the genomes of the symbiotic microbiota). The latter is made up of the genes of a variety of microbial communities that persist over time and are not eliminated by natural selection. Therefore, the holobiome can also be considered as the genomic reflection of the complex network of symbiotic interactions that link an individual member of a given taxon with its associated microbiome. Eukaryotic individuals can be analysed as coevolved, tightly integrated, prokaryotic communities; in this view, natural selection acts on the holobiont as if it were an integrated unit.

Recently I discovered another similar attempt, albeit rather shy. I mean the article by Kikvidze and Callaway (2009), wherein the authors tried to use ecological facilitation theory (see Chap. 5) to explain the occurrence of the first cells, eukaryotes, multicellular organisms, and societies. In my opinion, this article contributes quite well to the understanding of symbiogenesis.

This, however, is almost complete list of authors that I have managed to find. It is not easy to explain why so few scholars have chosen to follow Darwin's path from its beginning to the very end when they could benefit from the well-thought methodology, as well as new facts and concepts. I suspect that one of the reasons could be almost complete lack of interest in any scientific methodology among biologists. Most of them seem to be convinced that there is no need to reinvent the wheel and they can happily work with the methods used by chemists and physicists. Those sharing this attitude might think that chemists and physicists only recognise the inductive approach and all of them are inductivists and followers of F. Bacon (1561–1626). Bacon is well known as the author of inductive approach which he obviously preferred. He asserted that a human mind needed lead and weights, rather than wings, to prevent it from any leap or flight.

A possible reproach that the living world is much more complex than the non-living one and therefore not suitable for the deductive method is well countered by Murray (2001) who says that it has been rendered complex by us, biologists, through the application of inadequate methods. Some ecologists and evolutionary biologists seem to support this position (e.g. Lekevičius, 1986, 2002, 2011; Lawton, 1999). Besides, methodology, and theory-building methods in particular, is often considered the realm of philosophers, rather than biologists. The story of creation of the theory of natural selection presented above unsettles many biologists. They say 'No, Mr. Lekevičius, that cannot be true, you must be inventing things; even Darwin himself apparently wrote that he had produced his theory based on facts and by way of induction'. Deep inside, I have my own response to that: 'Jeez, why are you all so obsessed about the final result and conclusions, while paying so little attention to the method that has led to these conclusions? Look, the most sensitive and vulnerable to criticism part of any academic research is its methodology, not the results. Especially bearing in mind that the theory of evolution is the fundament of all life sciences'.

Finally, Darwin's methodology is not being emphasised and duplicated for yet another reason: the lack of need for all-encompassing synthesis of available knowledge, the need which was still obvious just half a century ago, and which modern physicists seem to have retained. Why so? The scope and the objectives of this monograph hold me back from engaging into deeper discussion on that subject, as it is way too broad (for more on methodology see Lekevičius, 2013).

Frankly, we, biologists, have always had problems with methodology. Biologists of the second half of the nineteenth century were not fond of the experimental method and tended to ignore it. According to them, let physicists and chemists muddle along with test-tubes and flasks, while biologists should take a different approach: all they must do is observe nature and draw some cautious conclusions. Eventually, C. Bernard, L. Pasteur, and G. Mendel broke this tradition, albeit not immediately. Together with experiments, mathematical methods and systemology have become more widespread. However, despite all historical lessons, deduction as a strategy for the development of new comprehensive biological theories is avoided up till now and is not even considered as an option (Murray, 2001). Darwin might have filled this gap unintentionally, but even that is not common knowledge. Many of us might not even be able to distinguish between the deduction method and the axiomatic one (proof of theorems), which admittedly has limited application in biology.

Physicists have made one procedure of the deductive method especially popular. This is logical, or thought, experiment, also referred to as mental modelling, because it is based on an attempt to simulate the situation 'What would happen, if…'. Such experiments were successfully used by Galileo, Kepler, Newton, Einstein, Bohr, and many other great physicists, and they contributed to the development of well-known physical theories. This method is also applied in mathematics, sociology, linguistics, and many other sciences. In general, thought experiments are used in circumstances where a real experiment would be very difficult to set up or is completely impossible. Besides, just as physical experiments, they are used to identify the factors of interest from among a much broader group and thus simplify the real world making it easier to understand. The heuristic value of these experiments should not be questioned. Quite often a thought experiment involves (again, for heuristic and simplification purposes) concepts representing imaginary, non-existent objects, or their qualities. Their examples could be a 'point' and 'line' in geometry, an 'absolute black body' and 'ideal gas' in physics, etc. Why are such concepts useful? According to Irvine (1991):

> … thought experiments have the advantage of being able to reach conclusions regarding hypothetical ideal environments (frictionless surfaces, perfectly elastic bodies, situations of absolute zero, and the like), thereby helping to identify and distinguish between nonaccidental variations between variables and parameter sets. By omitting from consideration particular variables (such as the friction of a surface) the dependence or independence of other variables may become clearer. Once this occurs, the underlying laws can then be applied profitably to more complex, real-world situations.

All that should be applicable not only in physics but also in biology. Especially, in the evolution of life. The theory of evolution is simply unthinkable without thought experiments or other instruments of deductive reasoning (Wächtershäuser, 1997). It would not be a grave mistake to consider Darwin's theory an example of a successful thought experiment. After all, he started with the thought of what would the world look like, if variability, heritability, struggle for existence, and selection had accompanied life from its very onset. Of course, this experiment would hardly meet modern adequacy requirements because it was not perceived as such by the author and was not properly defined.

I am also going to use the method of thought experiment. For example, I shall pose a question: what would happen if there were no diversity of genotypes and species, in other words, if populations and communities developed into a true super-organism. Or another one: what would happen if vacant niches provoked and stimulated evolution. If the conclusions drawn from such simulation will not contradict the available facts but rather provide a sufficiently convincing explanation, this will mean that my premise is correct, at least for the time-being. And on the contrary, if the conclusions contradict the empirical data or match it only under certain conditions, then these conditions should be analysed deeper, resulting in the adjustment of the initial premise. Some 40 years ago, I applied a thought experiment for the first time. After acquaintance with the ecosystem concept, I asked myself: what would happen if we successfully removed all decomposers from a tropical rain forest. The answer seemed very clear: the forest would die within a few weeks, because supplies of inorganic nutrients in these forests are usually rather limited.

To give a clearer picture of how these experiments are conducted, let me give another example. Let us imagine what would happen if the main 'goal' of life were to increase its total biomass on local as well as on global scale by all means available.

Here are the conclusions deduced from this premise (and from some postulates of the General Adaptation Theory, see Chap. 11):

1. In such case, life should be characterised by a tendency to increase biomass, for instance, by way of growth or propagation.
2. Both extensive (spatial spreading) and intensive (increase in density) growth of biomass should be observed in the process of ecological succession and evolution.
3. The intensive growth of biomass should be linked with improvement of its functional efficiency and shrinking costs of sustaining its viability.
4. To improve its functional efficiency and increase the total biomass, life could shift towards specialisation where individual structures focus on the performance of particular tasks.
5. In case of broad variation of abiotic environmental conditions on the planet, life could achieve higher and more stable total biomass levels not as a single superorganism, but by granting a certain degree of freedom to its specialised structures, i.e. through partial disintegration. In that case we would observe a number of spatially separated organisms, populations, or species; of course, they all should also demonstrate the trend to increase their biomass to the maximum possible level.

As it will be demonstrated later, these five conclusions are reliable and do not seem to contradict the available empirical data. Even more importantly, I think they provide quite convincing explanation why life on our planet has taken the course of partial disintegration into species and genotypes from its very onset.

Thus, I do not think that biologists or ecologists need lead and weights to constrain their fantasy and to make them cook their discoveries according to a proven recipe. On the contrary, it is about time for them to lose old clichés and stop being afraid to bring forward bold or sometimes even 'undoubtedly crazy' ideas. Their sceptical colleagues (who have never been lacking) will be able to sort things out. Even more so since we cannot boast of great achievements in biological theory. The majority of 'laws' or rules in evolutionary science and ecology are mere inductive generalisations describing the living nature rather than explaining it. Only the deductive approach can provide an answer to the question why the natural world is the way it is and not something else, and this is exactly the reason why hard sciences hold it in such high esteem (Lawton, 1999; Murray, 2001; Lekevičius, 2013). Inductivists, on the other hand, are not interested in unrealised situations, even though they might be worth discussing on theoretical level. In subsequent chapters, I shall present to my reader numerous examples of deductive reasoning and shall try to explain the workings of nature rather than merely describe them (see Fig. 2.3), even at the risk of being misunderstood. The scheme presented below might appear original, but it is not. It closely resembles the theory of deductivism presented by K. R. Popper (see Wächtershäuser, 1997).

One of the biggest problems of the theory of biology is the lack of crucial experiments aimed to verify that or another hypothesis (Platt, 1964). Even when somebody succeeds in carrying out such experiment, hordes of sceptics remain unconvinced and demand additional proof. Such stalemate situation may last for decades, as it happened with the issue of 'species diversity and stability of communities'. There have been hundreds of similar cases throughout the history of biology, although many of them are of less fundamental importance than the recent example. Anywhere within the theoretical field of modern biology one would encounter hypotheses waiting to be tested, and only a handful of them can already be considered proven theories. No wonder that the synthesis of biological knowledge has become a bit of a minefield.

To cope with this obstacle, I once (Lekevičius, 2002) devised a procedure which I named 'the principle of a meaningful sentence' or heuristic reliability. A reader would know that most words in any dictionary have more than one meaning, and some might have dozens of them. Nonetheless, we manage to understand each other pretty well. Under one condition, however: our sentences must be meaningful, i.e. maintain certain inner logic. This happens because in a meaningful sentence words make each other more specific, reducing the options of their meanings to the

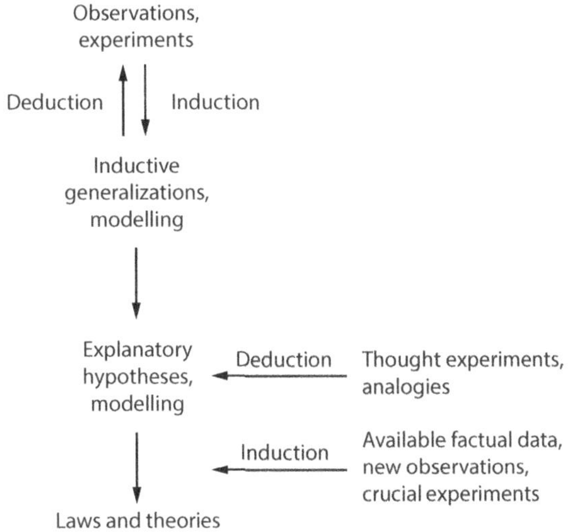

Fig. 2.3 A tentative sequence of procedures recommended to biologists seeking synthesis of knowledge. (From Lekevičius, 2013)

minimum necessary for a reasonable conversation. A similar result could be expected in case of hypotheses serving as 'words'.

Fortunately for biology and biologists, life as a phenomenon is objectively integral, thus the science dedicated to it should also be indivisible, compact, and gnoseologically simple. Indeed, the links between events and processes can always be found if they are sought after. In my understanding, the future general theory of biology is going to be a single, very meaningful, and logically perfect sentence, wherein the 'words' represented by hypotheses will not only gain their proper status but will shed most of their vagueness as they find their adequate meaning and place. Alas, for a whole century and more, we have been observing something different in real life: multiple fragments of the theory of biology, poorly linked and being split even further, so that integrating them into a single meaningful 'sentence' is becoming an ever more challenging task. Moreover, in the methodology of modern biology, there is not even a hint of a procedure explaining how to assemble a complete picture from multiple scattered fragments.

So, what exactly could the principle of a meaningful sentence offer to those still seeking an overarching synthesis? Let us start with the fact that there are numerous inductive generalisations, i.e. hypotheses to be found in literature, each focused on that or another aspect of the functioning or evolution of life. Moreover, many of these aspects or problems are often covered by more than one hypothetic solution, not necessarily clearly linked together. Despite that, the first step in the synthesis of knowledge requires to get acquainted with all these hypotheses with an open mind devoid of criticism and prejudice. By the way, the same approach is recommended in the well-known strong inference method (Platt, 1964). Let us also assume that, according to the empirical data, some of the hypotheses and solutions are more likely to be correct than wrong, although, as it often happens, none of them has a definitive proof. These hypotheses and solutions have been presented and reviewed by groups of highly specialised scientists, and none of them might have even thought about the relations between them, let alone more global synthesis of knowledge. I think that the principle of a meaningful sentence could be useful for the latter objective. I would follow the assumption that when a 'sentence' is really meaningful or free of logical inconsistencies, the fallacy index f ($0 \leq f \geq 1$, where 0 denotes complete truth and 1 – complete falsity) of the 'sentence' should acquire a lower value, as if the corresponding indices of the comprising 'words' were multiplied together. Even if some 'words' were less popular among experts ($f > 0.5$), they could still be included in the 'sentence', provided that the alternative hypotheses that had attracted much broader support would be even less compatible with the rest of the 'sentence'. Of course, I omitted real calculations in my exercises because the result would have been hopelessly subjective, and thus inaccurate. For me, it was enough to realise that, because of the links between 'words' (i.e. hypotheses), the uncertainty of each of them, as well as that of the general 'sentence', would tend to decrease, rather than increase.

Let me give a few examples.

Example 2.1

Nowadays, the so-called unified neutral theory of biodiversity (UNTB) whose author is Hubbell (2001) has gained a broad circle of supporters. This theory states that the niches of coexisting plants are identical, while species diversity and population sizes of these plants are determined by casual factors. Seen in isolation, this concept appears free of contradictions, some would even say that it resembles the truth ($f \approx 0{,}2$) – no wonder it has received so much attention and support from our colleagues (see Chap. 5). However, when this 'word' is placed in a broader context in attempt to produce a meaningful 'sentence' by adding also other well-known ecological concepts (e.g. Gause's principle) and non-questionable postulates of the theory of evolution, it simply does not work. Despite all efforts and combinations, the 'sentence' always includes internal contradiction. It seems that all previous conceptual assets accrued throughout decades contradict this effort of synthesis of knowledge, therefore either all these assets should be discarded, which would surely lead to a real disaster, or the newly coined 'word' should be rejected as incompatible with the broader theoretical context.

Example 2.2

The first 'word': Wild populations demonstrate rather high genotype diversity which is not just temporary.

The second 'word': The reaction norms encoded by different genotypes of the same population are somewhat different. Therefore, with environmental conditions changing in space and/or time, different genotypes alternate in getting the advantage in growth and reproduction.

The third 'word': Observations and experiments seem to suggest that such biodiversity might be supported by the diversity of environmental conditions.

The fourth 'word': Diversity of genotypes provides populations with additional functional stability under changing environmental conditions and maintains total biomass and productivity at a certain level.

The fifth 'word': Responses of a gene pool to the changing environment are accompanied by adjustments within the limits of individual reaction norms.

The sixth 'word': Gene pool and phenotypic reactions require some time; besides, the former take longer. Therefore, the environment may change again sooner than the population manages to adapt to the previous change. Thus, complete adaptation and optimal conditions seem to be a moving target rather than a stable state.

The generalising 'sentence' (which presumably contains less uncertainty than any separate 'word' listed above): In the course of day-to-day activities of populations, natural selection (differential reproduction of pre-adapted genotypes) and phenotypic plasticity not only ensure survival of these populations but also facilitate their adaptation to the ever-changing environmental conditions; phenotype plasticity makes population optima fitting more delicate than if driven by genotype diversity alone.

Example 2.3

The first 'word': There are reasons to believe that, over a few million years, a certain species that was among the first to settle on a volcanic island would have produced dozens of endemic species occupying different niches.

The second 'word': In the early stages of colonisation, vacant niches, i.e. resources without specific consumers, must have been abundant. Some of these niches might have met the needs of the newcomer species, while others not.

The third 'word': Populations of the same species that remained in their original habitat (on the continent) have not undergone any significant diversification over the same period.

The fourth 'word': Populations remaining in the original habitat have never lost their mutation and recombination capacity.

The generalising 'sentence' (which presumably contains less uncertainty than any separate 'word' listed above): Vacant niches must have been the main driving force of this adaptive radiation. They acted both as opportunities to evolve and as constraints setting the courses of that evolution.

Example 2.4

The first 'word': Earth ecosystems, especially those that share the same biome, demonstrate many common features: almost identical material cycles, similar production pyramids, similar sets of guilds, and large number of equivalent species. There are reasons to believe that organisational structure of ecosystems has not changed significantly from the late Palaeozoic. This is the phenomenon of functional convergence of ecosystems.

The second "word": Some ecosystems are formed by way of immigration (ecological succession), some others by way of immigration and rapid diversification (e.g. on remote volcanic islands). Yet another part, such as the first terrestrial ecosystems in the Palaeozoic, almost exclusively by way of evolution.

The third "word": Despite differences in mechanisms and speed, both modern primary succession and formation of terrestrial ecosystems in the Palaeozoic are characterised by similar phases: the first groups to settle in are photosynthetic and chemosynthetic organisms as well as decomposers, followed by organisms of the second trophic level and ultimate establishment of predators.

The fourth 'word': Similarities can be observed not only in phases but also in the developmental trends of the functional parameters of ecosystems: increase in species diversity and increase of the total biomass up to a certain ceiling level (by way of extensive and intensive growth).

The fifth 'word': This self-arrangement of ecosystems proceeds with acceleration, just like in case of autocatalysis, because photosynthetic and chemosynthetic organisms produce detritus necessary for the settlement of decomposers or their evolution, while also creating vacant niches for future 'herbivores', and the latter eventually provide niches for future predators.

The generalising 'sentence' (which presumably contains less uncertainty than any separate 'word' listed above): Both the process of succession and self-organisation of communities and ecosystems by way of evolution are subject to the same organisational constraints.

Finally, a couple of words on the modelling approach used in this monograph. Positivistic science demands for mathematical expression of the results produced by studying nature. In my opinion, however, mathematical modelling has made far less contribution to ecology and the theory of evolution than it was expected to merely some three or four decades ago. I am not going to repeat myself, as I have sufficiently covered this topic elsewhere (Lekevičius, 2013). Let me only quote the opinion of Krebs (1988), one of the most experienced ecologists and methodologists, on this matter:

> Mathematical models are not a means of discovery and cannot be used to reject logically impossible ideas in the real world (...) Nor are mathematical models useful for defining concepts operationally in ecology, and I doubt if models ever produce new principles in ecology.

In one of his programmatic articles, Krebs (2006) regretfully admits that any meticulous observer would notice obvious conceptual and methodological deficiency of most ecological models and their detachment from reality; rather than providing solutions to identified problems, they tend to create confusion among the academic community.

Harsh as it may be, this criticism is somewhat justified. The thing is that biological objects are indeed 'hopelessly' diverse, different from one another, and therefore they are not suitable for the traditional machinery of mathematics. Lane (1985) has made a very accurate observation:

> Ecosystems (...) are neither small nor large number systems. They contain too many components to be treated analytically and too few for statistical analysis. A fish and a phytoplankton cell cannot be averaged like ideal gas molecules ...

Early in my career, I have had experience with mathematical modelling of ecosystems and used to communicate a lot with outstanding Lithuanian and Russian mathematicians. After all that experience, I ended up with conclusions similar to those of C. J. Krebs. Frankly, even modern biological literature is full of examples where quantitative descriptions are used without qualitative clarity. It would not be a misstatement to say that the prevailing attitude is not even positivistic but, rather, scientistic. Are they attempts to produce the impression of 'deepness' or 'high scientific level'? Or is it done with the expectation that a mathematical model will help to conceal the conceptual feebleness of its author? I am not sure about the motives, but I am sure of something else: no model can be better than the data set it is based on, no model can fill our gaps in knowledge, correct our errors and advise how to avoid them. Empirical discovery will always remain the ultimate criterion of truth. Wilson (2013), whom I quote in numerous instances, had made the following statement in his provocative article 'Great scientist ≠ good at math', which I find quite justified:

> Far more important (than mathematical fluency – E. L.) ... is the ability to form concepts, during which the researcher conjures images and processes by intuition.

By the way, A. Einstein has also emphasised on various occasions that he credited his imagination for his discoveries, rather than his mathematical skills, which, as we know, were not exceptional. However, *Quod licet Iovi, non licet bovi*. These words written by Wilson, perhaps the most outstanding contemporary biologist, produced thousands of new opponents, even though he had never had any shortage of them. I was impressed by something else, namely by his courage (as well as that of Krebs). In the era of the almighty conjuncture dwelling on the 'physics envy', throwing such words at the crowd is comparable to a confession in homicide or uttering the truth about 'the emperor's new clothes'. Regardless of the ultimate outcome of this dispute, an exciting intrigue of the meta-theory level seems to have appeared in biology, making our lives more interesting.

This is why I am going to almost completely avoid the results produced by mathematical models and would preferably rely on empirical data in this monograph. Nonetheless, I support the positivist position that science should be precise and cannot be limited to verbal descriptions only. Therefore, my reader is going to find not only rather strict definitions of the key concepts but also a number of examples of conceptual modelling. Construction of these models does not require any particular skills, they make it easy to present thoughts and ideas, they are well suited for both illustrating the links between already known objects, phenomena, or processes, and formulating new hypotheses. They are appropriate both for synthesis and analysis. Most importantly, there is no level of complexity that a conceptual model could not cover; it all depends on the creativity of the researcher and the level of abstraction. This quality makes conceptual modelling very different from mathematical modelling. In case of the latter, usually not the model is adapted to nature, but, on the contrary, nature is simplified to the level suitable for mathematical processing, thus rendering the entire effort useless (Lekevičius, 2013).

Conceptual modelling may be especially handy in cases when the data about the object, phenomenon, or process to be simulated are scarce or unreliable. I therefore stick to the opinion that any research into a new or challenging problem should start with a hypothesis that can be transformed into a conceptual model, i.e. with the merest description. Only later, in case of success, a more detailed quantitative description should follow, if necessary and achievable.

References

Carey, T. V. (1998). The invisible hand of natural selection, and *vice versa*. *Biology and Philosophy, 13*, 427–442.

Corning, P. A. (2005). *Holistic Darwinism: Synergy, cybernetics, and bioeconomics of evolution*. University of Chicago Press.

Guerrero, R., Margulis, L., & Berlanga, M. (2013). Symbiogenesis: The holobiont as a unit of evolution. *International Microbiology, 16*, 133–143.

Hubbell, S. P. (2001). *The unified neutral theory of biodiversity and biogeography*. Princeton University Press.

Irvine, A. D. (1991). Thought experiments in scientific reasoning. In T. Horovitz & G. J. Massey (Eds.), *Thought experiments in science and philosophy* (pp. 149–165). Rowman & Littlefield Inc..
Kikvidze, Z., & Callaway, R. M. (2009). Ecological facilitation may drive major evolutionary transitions. *BioScience, 59*, 399–404.
Krebs, C. J. (1988). The experimental approach to rodent population dynamics. *Oikos, 52*, 143–149.
Krebs, C. J. (2006). Ecology after 100 years: Progress and pseudo-progress. *New Zealand Journal of Ecology, 30*, 3–11.
Kropotkin, P. (1902). *Mutual aid: A factor of evolution*. Heinemann.
Lane, P. A. (1985). A food web approach to mutualism in lake communities. In D. H. Boucher (Ed.), *The biology of mutualism: Ecology and evolution* (pp. 344–374). Oxford University Press.
Lawton, J. H. (1999). Are there general laws in ecology? *Oikos, 84*, 177–192.
Lekevičius E. (Лекявичюс Э.). (1985). Explanations in biology: An attempt to synthesize functional and causal approaches. In R. Pavilionis (Ed.), *Scientific knowledge: Systemic aspect* (pp. 119–135). The Academy of Sciences of the Lithuanian SSR. (in Russian).
Lekevičius E. (Лекявичюс Э.). (1986). *Elements of a general adaptation theory*. Mokslas Publ. (in Russsian).
Lekevičius, E. (2002). *The origin of ecosystems by means of natural selection*. Institute of Ecology (Vilnius).
Lekevičius E. (Лекявичюс Э.). (2009). On some analogies between ecosystem evolution and development of economy: From A. Smith and C. Darwin to the latest ideas. In L. E. Grinin, A. V. Markov, & A. V. Korotaev (Eds.), *Evolution: Cosmic, biological, and social* (pp. 226–260). Moscow. (in Russian).
Lekevičius, E. (2011). Ecological Darwinism or preliminary answers to some crucial though seldom asked questions. In L. E. Grinin, A. V. Korotaev, R. L. Carneiro, & F. Spier (Eds.), *Evolution: Cosmic, biological, and social* (pp. 101–121). Uchitel Publishing House.
Lekevičius, E. (2013). *Ecology: From individuals to the biosphere*. Vilnius University. (in Lithuanian).
Looijen, R. C. (1998). *Holism and reductionism in biology and ecology*. Dissertation. University of Groningen. http://dissertations.ub.rug.nl/faculties/fil/1998/r.c.looijen
Margulis, L., & Sagan, D. (1986). *Microcosmos: Four billion years of evolution from our microbial ancestors*. University of California Press.
Mayr, E. (1968). Cause and effect in biology. In C. H. Waddington (Ed.), *Towards a theoretical biology. Vol. 1: Prolegomena. An IUBS symposium* (pp. 42–54). Edinburgh University Press.
Murray, B. G. (2001). Are ecological and evolutionary theories scientific? *Biological Reviews, 76*, 255–289.
Platt, J. R. (1964). Strong inference: Certain systematic methods of scientific thinking may produce much more rapid progress than others. *Science, 146*, 347–353.
Schweber, S. S. (1977). The origin of origin revisited. *Journal of the History of Biology, 10*, 229–316.
Vermeij, G. J. (2004). *Nature: An economic history*. Prinston University Press.
Wächtershäuser, G. (1997). The origin of life and its methodological challenge. *Journal of Theoretical Biology, 187*, 483–494.
Walker, J. C. G. (1980). Atmospheric constraints on the evolution of metabolism. *Origins of Life, 10*, 93–104.
Wilson, E. O. (2013). Great scientist ≠ good math. *Notices of the American Mathematical Society, 60(7)*, 837–838.

Chapter 3
Species Do Not Just 'Struggle'

First, a few words on the concept of 'biological diversity'. For the purposes of this monograph, it should be interpreted as the diversity of molecules, genes, subcellular structures, cells, tissues, organs, genotypes, species, and ecosystems on planet Earth, although my focus will be on the last three types of diversity. I am also going to emphasize two types of species diversity – α and β diversity. The first one refers to species diversity within a local community, while the second characterizes the differences in species composition of neighbouring communities (Whittaker, 1977). Nevertheless, both these concepts have rather limited application because usually (although not necessarily) there are no clear boundaries between one local community and another neighbouring community. Species diversity takes into account both species richness or the number of species, and species relative abundance or equitability. In most cases within this monograph, I am going to use the concept of species diversity in its narrow sense, as a synonym of species richness.

At present, biologists have described over 1.5 million species, but there are no doubts that their actual number is much higher. Some 5000–10,000 new species are discovered annually, most of them being tropical insects. Wilson (1992, 2003), already mentioned earlier, is convinced that our Earth is home to at least a few tens of millions of species. Arthropods alone should account for a few or perhaps even a dozen millions of them. These numbers, however, might as well be a bit exaggerated. Besides, species-defining criteria are not completely established in some fields, such as microbiology, therefore we should hardly expect more precise estimates in the near future. Nevertheless, even by the most conservative estimates, this diversity is astonishing.

Let us start analysing the type of interactions between organisms. Some ecologists consider populations and communities to be true systems, just as ecosystems (e.g. Odum, 1971; Patten & Odum, 1981; Berryman & Kindlmann, 2008). According to Odum (1971, p. 8), 'any unit that includes all of the organisms (i.e. the "community") in a given area interacting with the physical environment, so that a flow of energy leads to clearly defined trophic structure, biotic diversity, and material cycles (i.e. exchange of materials between living and non-living parts) within the system is an ecological system or ecosystem'.

The phrase 'a system is more than the total of its composite parts', often reiterated by ecologists, means that certain characteristics are inherent in the entirety only, but not in its individual components. These characteristics are called emergent. They include population size and density, genotype diversity, age and sex

E. Lekevičius, *Biodiversity*, https://doi.org/10.1007/978-3-031-11582-0_3

structure, etc. Correspondingly, a community and an ecosystem in their entirety are characterised by species diversity, domination indexes, and some other parameters not applicable to individual species. Even more important are functional emergent characteristics. In case of a population, this is first of all reproduction, especially when we speak about sexually reproducing species where one partner is unable to perform reproductive function without another. On the level of ecosystem, we have a nutrient cycle and the associated energy flow – these are also emergent characteristics beyond any doubt because they apply neither to individual species nor to abiotic media.

One may get the impression that there is little to add. Big deal – some emergent characteristics! They are inherent even to non-living objects. For instance, the properties of sodium chloride cannot be derived a priori from the data on the properties of sodium and chlorine atoms. A stone can crack somebody's head open, while a single molecule entrapped within that stone does not have this capacity. Emergentism appears a rather trivial fact in this context, with hardly any deeper meaning. Indeed, ecologists often stop at this point and do not dare to move any further.

Which is a pity, because the emergent characteristics of life also indicate division of labour, or functional hierarchy, and the latter can by no means be called a trivial fact. Let us check what the creators of the ecosystem concept write about division of labour. According to Odum (1971), the first principle of ecology should describe the types of interactions between organisms and their interactions with abiotic environment. His own formulation of this principle reads as follows:

> The individual organism, for example, cannot survive for long without its population, any more than the organ would be able to survive for long as a self-perpetuating unit without its organism. Similarly, the community cannot exist without the cycling of materials and the flow of energy in the ecosystem (Odum & Barret, 2005; p. 7).

In other words, neither a single individual of any species nor a species itself are functionally independent. Odum (1971) clarifies that all organisms of an ecological community can be classified into three major groups or functional kingdoms of nature: producers, macroconsumers, and microconsumers or saprotrophs. Both macro- and microconsumers are not independent in terms of functioning, they depend on producers that supply nutrients and energy. Producers, however, are not independent either, at least in terms of materials, if not energy. Should microconsumers not recycle inorganic nutrients as the ultimate products of their metabolism, producers would soon perish thus drawing to extinction also macroconsumers, or biophages (herbivores, predators, and parasites). The authors of the ecosystem concept have persistently emphasised the approximate balance between the production of organic material and its decomposition, including the loss and uptake of inorganic nutrients. Life could hardly exist without such a balance. After all, even a slight surplus in nutrient loss would have turned current ecosystems into lifeless desserts during millennia.

Thus, the relationship between decomposers (≈ microconsumers ≈ detritivores ≈ detritophages ≈ saprovores ≈ saprophages ≈ saprotrophs) and producers can be considered as an indirect although obligatory mutualism (see also Lekevičius, 2000,

2002; Loreau, 2010; Veldhuis et al., 2018). One may say that only an ecosystem has the capacity of staying viable for indefinite periods of time:

> The ecosystem is the first unit in the ecological hierarchy that is complete – that has all the components (biological and physical) necessary for survival (Odum & Barret, 2005, p. 18).

The concept of interaction between the functional kingdoms of living nature, supported by these ecologists, could be presented in a simple model (Fig. 3.1):

This, according to Odum (1971), does not mean that elimination of a single species from the community should result in a catastrophe for the entire community. As I would rephrase it, no species is not only a sufficient but also a necessary condition for the existence of a local nutrient cycle. Ecological communities are 'insured' against such catastrophes in numerous ways, and I shall discuss them a little later.

The ecosystem concept suggests that nothing can bind all the organisms of a community together stronger than nutrient cycles wherein 'primitive' life forms play a key role. There are over 20 chemical elements absolutely necessary for growth and reproduction, or at least normal functioning of living organisms. They are grouped into macronutrients (C, O, H, N, P, Ca, Mg, S, Cl) needed by organisms in relatively high quantities, and micronutrients (Fe, Mn, Mo, Cu, Si, Na, I, etc.) sufficient in much smaller quantities. All of them are necessary, although different organisms have different needs. In the process of anabolism, chemical elements are transferred from their inorganic form into the organic one, while catabolism (respiration) brings them back from the organic form into the inorganic. Life can only function under the condition that all these 20 elements are recycled in a harmonised way: their cycles work either all together, or none (Liebig's law of minimum).

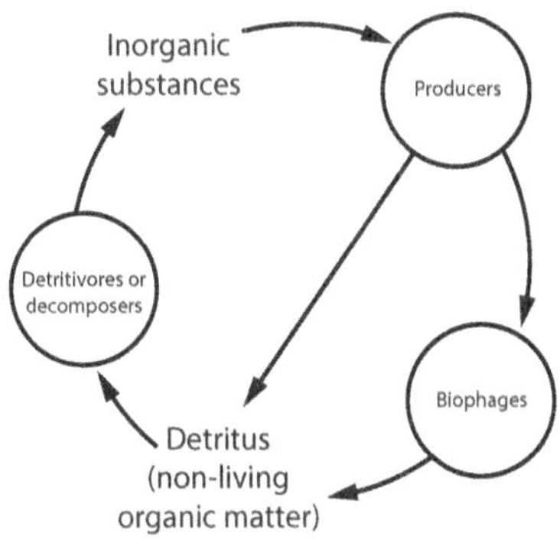

Fig. 3.1 An extremely simplified model of an ecosystem. The arrows indicate the directions of transfer and transformations of materials between the building blocks of the ecosystem. The scheme is aimed at maximum simplification of the real picture which, unfortunately, resulted in some distortions. One may notice that some important links are missing, even though they might not be of the primary importance. Still, additional arrows pointing from producers and biophages (herbivores, predators, and parasites) towards inorganic substances could be useful – after all, these organisms respire and thus produce biogenic substances. Another arrow could be drawn from decomposers towards detritus (decomposers also die and become part of detritus). Finally, biophages feed not only on producers and one another, but also on decomposers. One way or another, these additional arrows would not change the overall picture. (From Lekevičius 2002)

Therefore, interdependence of different life forms is equally determined by their dependence on one of these cycles and by the interdependence of different cycles.

Pattee (1968), a renowned expert of theoretical biology, has once made a memorable statement that also perfectly illustrates the attitude of most ecologists of that period (p. 219):

> There is no living unit which can be considered 'living' without reference to the external environment. (...) Biologists should emphasize over and over that "living" is unavoidably a total ecosystem property and not the property of an isolated collection of macromolecules. It seems to me that the central question of the origin of life is not, "Which comes first, DNA or proteins?", but rather "What is the simplest possible ecosystem?"

Consequently, the paradox of primitive life forms (see Chap. 1) is not seen as a paradox by the ecologists who are well familiar with the ideas of L. Pasteur, S. N. Vinogradsky and E. Odum. Are you asking why primitive organisms still exist? Because most of them not only avoid 'struggle' with more modern and advanced organisms, but even cooperate with them, thus making life possible. To put it in other words, primitive organisms occupy the niches wherein their adaptedness is just as good as that of advanced life forms in their different niches. Of course, this is a merely functional, therefore incomplete, explanation.

Let us proceed with the issue of the division of labour and mutual aid, however. Ecologists mark mutualistic relationship with a double plus mark (++) indicating mutual benefit. Relationship between the exploited organisms and the exploiters (plants and grazers, prey and predators, hosts and parasites) is symbolized by a minus and plus sign (−+), suggesting that one partner is losing while the other is gaining from this relationship. There are opinions that the minus should not be interpreted as unconditional (Berryman & Kindlmann, 2008; Krebs, 2009). If the partners have been in contact for sufficiently long period, they start to exploit the living resources in a sustainable or moderate manner due to coevolution (Utida, 1957; Pimentel, 1968); too aggressive or too virulent genotypes in the exploiter population are eliminated by natural selection, although too 'merciful' ones are not tolerated either. This can be understood, because the consumers that destroy their feeding resources cause harm to themselves (a negative feedback). For instance, if a parasite is moderately aggressive, the host population is encouraged to grow, and a dense population ensures better spreading conditions for the pathogenic organisms than a sparse one. Incredible as it may sound, but parasites, just as other exploiters, are 'interested' not in extermination of the exploited populations but rather in their increase. Another common knowledge is that predators contribute to the selection of their prey by killing many individuals infected with parasites or suffering from diseases. Thus, indirectly, parasites often cause the death of their host, while predators prevent parasites from spreading. By the way, a prey or host species that is rid of predators or parasites and thus allowed to sharply increase its numbers may often deplete its own food resources to unsustainable levels. Admittedly, 'struggle for existence' in nature appears cruel; the sights of blood, suffering and death inevitably provoke our compassion and call for a minus sign. However, seen from the population viewpoint, rather than from that of an individual, this phenomenon appears somewhat different and boils down to the regulatory role of the exploiters (Kontrimavičius, 1982; Valkiūnas, 2001). I might have presented a little too simplistic picture of the actual state of affairs in nature because some population ecology experts (e.g. Begon et al., 2006; Krebs, 2009) see it as somewhat more complex and less straightforward.

Finally, we should bear in mind that, according to the same 'plus/minus' assessment system, all the even as well as all the uneven trophic levels do not struggle among themselves but cooperate, and their relationship should be labelled with a double plus sign (++). Indeed, wolves are the best friends of plants, aren't they? A good example of such impact could be the relatively well-studied effect of the extermination and reintroduction of wolves on the Yellowstone National Park communities (Smith et al., 2003; Robbins, 2004). The reintroduction resulted in significant drop of the herbivore populations but also in recovery of the plant communities. This phenomenon is sometimes called a trophic cascade. Such indirect mutualism is observed in all food chains of biophages (herbivores, predators, and parasites), and it undoubtedly stabilises the populations of interacting species thus supporting species diversity (Shurin et al., 2002; Duffy et al., 2007).

However, examples of direct mutualism (++) are equally abundant. These are, among others, a mycorrhiza which is an association between most plants and fungi, also plant associations with nitrogen-fixing bacteria, links between the flowering plants and their pollinators and distributors of seeds, the union between a fungus and an alga (sometimes – cyanobacterium) in lichens, the role of symbiotic microorganisms in cellulose and lignin digestion, association between zooxanthellae and corals, and many more. In general, mutualistic relationship is so widespread in nature that it would be difficult to find organisms without such connections (Begon et al., 2006; Taylor et al., 2009; Guimarães et al., 2016). Some time ago we had the impression that cooperation is quite rare among microorganisms, however, recently this opinion had to be revised (Schink, 2002) – there is more and more evidence that syntrophic, metabiotic, and other interdependence forms are much more common than it has been considered before.

Thus, we clarified that nutrient cycle and flow of energy is an emergent quality of a community, which is not typical to any individual species of that community and which is produced by the common activity of all its organisms. In other words, a common function (a nutrient cycle and the associated energy flow) is split into smaller functions performed by separate 'functional kingdoms'. Presumably, these large functions are themselves divided into even smaller ones represented by individual guilds and populations (species). All that will further be discussed in other chapters. Nevertheless, neither ecological communities, nor biosphere as the totality of communities and ecosystems cannot be considered super-organisms because their constituent parts (individuals and populations) retain significant freedom of action which they can – and indeed do – use for their own 'goals', i.e. they reproduce and spread on the slightest occasion regardless of the long-term impact on the viability of the entire system. All species are subject to this syndrome which I call malicious growth. The biggest and most important collision in the living nature might be this incompatibility between the 'goals' of individuals and species (populations), on the one hand, and their ecological community, on the other hand (Lekevičius, 1986a, b).

Somewhere back in 1967, A. Koestler, a famous novelist, essayist, and promoter of the Systems theory, gave a rather precise characterisation of the relationship between the totality and its parts in living systems by using the symbol of two-faced Janus (see Koestler, 1979). One of his faces looking towards the higher organisational level beams with humbleness while the other one turned towards the lower level represents the face of a despot pursuing only his own interests. As I am writing these words, I see a mental image of a petite merchant who meets his customers with a humble smile – he seeks only his profit but to earn it he must please the customers in the best possible way. In other words, Koestler stated that parts of a system, or subsystems, are both autonomous and dependent on the total at the same time. This circumstance – a relatively generous freedom of action – should be interpreted as a breach or weakening of the functional hierarchy, but not as its complete denial because the freedom enjoyed by organs and organisms is delimited by their functional dependency. Admittedly, Koestler failed to provide an unambiguous answer whether this hierarchy reaches the levels of population and community. However, there is another reason for mentioning this thinker: it was he who, in 1968, organised the Alpbach symposium *Beyond Reductionism – New Perspectives in the Life Sciences*. The event was attended by a surprisingly diverse company, including the world-famous authorities in science: L. von Bertalanffy, C. H. Waddington, P. A. Weiss, W. H. Thorpe, V. E. Frankl, F. A. Hayek, J. Piaget, etc. One of the focus points at the symposium was the threat posed by the spreading reductionism to biology, psychiatry, and other sciences. Unfortunately, this event seems to have had little impact on the subsequent development of the mainstream science.

Has the ecosystem concept emphasising the dependency of organisms in their functioning been denied during the last several decades? Apparently not, at least not to my knowledge. Even those ecologists who are traditionally considered supporters of population-focused approach (e.g. Begon et al., 2006) admit that the above-mentioned functional hierarchy or division of labour does exist, and no life could exist without decomposers. To be fully objective, I should mention one more thing: the fact that decomposers are necessary for supporting life on Earth was stated already by L. Pasteur and S. N. Vinogradsky, two great microbiologists of nineteenth and early twentieth centuries. Of course, back then these declarations must have been well-ahead of their time, therefore they failed to reach our consciousness. But what kept us from getting alert when the same words were repeated just a few decades ago by E. Odum? Indeed, if they are right a natural conclusion is that 'struggle for existence' is not – and might have never been – the only principle ruling the functioning of life and its evolution. In this context, species become not only self-sufficient goals but also instruments, because each of them makes a certain input into the maintenance of nutrient cycles. They cannot be self-sufficient goals already for the sole reason that no species can function independently.

The conclusion that individuals and species are not independent units and that functional hierarchy may reach as far as the community and ecosystem levels might not appear very surprising to most scholars. I am sure I could easily convince any, even a mediocre, biologist that it should not be questioned. On the other hand, I have an impression that it has never been perceived as a fundamental fact, perhaps even the most important discovery of the twentieth century. Hence there were no conditions for the understanding that this conclusion should serve as the basis for the future general theory of biology. A theory which would consolidate all hypotheses and theories produced or being developed by ecologists, physiologists, biochemists, and geneticists into a single system, which would merge individual 'words' into a single 'sentence' (see Chap. 2), free of internal contradictions and easy to understand. As stated, nowadays most biologists do not seem to feel an urge to develop an overarching theory, and they might have their reasons – they simply fail to properly appreciate or notice the functional interdependence of individuals and species. On the other hand, looking retrospectively, I clearly realise the following: we ought not to have shattered to pieces the whole produced by nature itself and then wandered around picking up the shards. Analysis is a serious and useful undertaking, but there is time for breaking pots and gathering the shards, and there is time for kneading clay and throwing new ones. Frankly, taking a closer look at that 'pot' before smashing it would have been much wiser.

Besides, in my opinion, as soon as we understood that individuals and species are not independent, we ought to have started adjusting the theory of evolution and constructed an even more meaningful 'sentence'. Such adjustment might have eliminated at least part of our paradoxes and provided an explanation of the occurrence of the present species diversity. Unfortunately, this ecological message has had no impact on the official version of the theory of evolution.

Another equally important fact has also produced no impact there. I mean the evidence of common features shared by all local communities and ecosystems, which I named a functional convergence of ecosystems not so long ago (Lekevičius, 2000, 2002). The evidence itself is rather old: more than six decades ago, Odum (1953) presented the idea that all local ecosystems, including land, marine and freshwater, share common features. One of them is the presence of all the three 'functional kingdoms of nature' and their interaction in creating a local energy flow and nutrient cycle. Despite the fact that Earth's ecosystems vary significantly in their species composition, their functional portraits are similar. Here are the words of a famous scholar Darlington taken from his book *Zoogeography: The Geographical Distribution of Animals* (1957, pp. 552–553):

> ...the most significant thing of all is the general level and balance of past and present faunas. Throughout the recorded history of vertebrates, whenever the record is good enough, the world as a whole and each main part of it has been inhabited by a vertebrate fauna which has been reasonably constant in size and adaptive structure. Neither the world nor any main part of it has been overfull of animals in one epoch and empty in the next, and no great ecological roles have been long unfilled. There have always been (except perhaps for very short periods of time) herbivores and carnivores, large and small forms, and a variety of different minor adaptations, all in reasonable proportion to each other. Existing faunas show the same balance. Every continent has a fauna reasonably proportionate to its area and climate, and each main fauna has a reasonable proportion of herbivores, carnivores, etc. This cannot be due to chance. Something holds the size and composition of vertebrate faunas

> within limits in spite of continual changes and successions in separate phylogenetic groups. Only competition can do this, and to do it competition must be a fundamental, ever-present force.

Later, it became clear (Odum, 1971), that these similarities extend into chemical transformations or, in other words, nutrient cycles, as well as into ecological pyramids, especially the production pyramids (often erroneously labelled 'energy pyramids'). Moreover, it was discovered that the shape of production pyramids and the number of trophic levels are almost independent from the primary production, i.e. the width of the pyramid bases which can differ tenfold or more. This is easily understood, because the shape of a production pyramid is determined by the laws of thermodynamics which are independent of the local variations. These facts were so important that Cody (1974) could not help but write:

> Such a degree of convergence gives reason to believe that there is a single optimal way of dividing up the resources.

This calls for an explanation. Thermodynamics does not allow for the pyramids to become rectangles, while biology does not allow gross production efficiency, or trophic level transfer efficiency (*P/I*, where *P* is production or biomass increase per time unit and *I* is the mass of food consumed per time unit) to exceed 0.5 or to decrease below 0.01, which, taken together, results in fluctuation in the proximity of 0.1 (the 10% rule). It is worth noting that this percentage is somewhat higher in aquatic ecosystems because of the higher efficiency, compared with terrestrial ones. This difference is not essential albeit somewhat peculiar – aquatic ecosystems support more trophic levels, up to 7 or even 8, whereas terrestrial ecosystems might have 1 or 2 levels less.

Note. It seems that lately ecologists tend to drift away from the concepts of production (energy) pyramid and trophic levels, as well as the 10% rule. Not without reason, they state (e.g. Polis & Strong, 1996) that in nature we mostly observe omnivory and food web dynamics, rather than linear food chains. Nonetheless, I shall insist that these concepts and the 10% rule are useful, albeit coarse, generalisations. I believe that abstraction aimed at filtering the less important details as well as idealisation are justified methods of logical simplification which have been and will remain useful tools in tackling the complexities of the surrounding world.

There are more similarities. For instance, tropical rainforests along the equator (in South America, Africa, and Southeast Asia) practically do not have common tree genera, however their growth forms are strikingly similar (Richards, 1957; Ewel & Bigelow, 1996): buttress and stilt roots, large broad leaves, numerous lianas and epiphytes on their trunks and branches. The soil in these forests contains relatively low quantities of inorganic nutrients, whereas the number of plant species may be very high (provided the annual precipitation is comparable; in the areas with significantly lower precipitation, such as inland Africa, the number of plant species is lower). Some authors (e.g. Reich et al., 1997) distinguish also ecophysiological convergence patterns even between plant communities assigned to different biomes of the world, ranging from tropics to deserts and alpine tundra. On top of that, we could add thousands of ecological equivalents – not closely related but convergent species found in different geographical regions and occupying similar ecological niches (see Odum, 1971; Pianka, 2000). For instance, many equivalents are observed among the nectar-feeding birds: Hawaiian honeycreepers (tribe Drepanidini, family Fringillidae), honeyeaters (family Meliphagidae; New Guinea, Australia, New Zealand), sunbirds (family Nectariniidae; Africa, Middle East, South Asia, Australia), and hummingbirds (family Trochilidae, both Americas).

During the 1970s, convergence at the levels of species, communities, and ecosystems has become generally accepted and found its place in many textbooks. The conviction was growing that even in distant geological past, e.g. in the Mesozoic, the structure of communities already attained the form very close to the present one (see Odum, 1971; Wilson et al., 1973; Richardson, 1977):

> Again and again, on different continents and islands and in different geological times, evolution has created ecological communities that resemble each other in their broadest features. The origins of the species that make up these separated communities are often drastically different, but the final products are usually remarkably similar (Wilson et al., 1973, p. 852)

Thus, this convergence has gained additional invariant features in our consciousness. The authors of the ecosystem concept were incredibly lucky – had these common characteristics not been discovered, then many individual concepts would have to be created instead of one. Unfortunately, they emphasised these facts too seldom and failed to grant this convergence a proper specific status. Could this be because they, as ecologists, were not very keen on seeking somewhat deeper evolutionary explanation, and thus leaving the very fact to appear suspended in the void? Or, maybe, they were perceiving this convergence as a trivial fact? I do not see it that way, especially bearing in mind the well-known fact that ecosystems can be assembled in two rather different ways: through immigration and/or evolution. The ways might be different, but the result is similar – what could that mean? Unfortunately, this question either was never posed or simply failed to attract any significant interest.

Meanwhile, the evolutionary biologists apparently were not interested in this topic or missed it out altogether. Eventually, as the holistic paradigm was being gradually replaced with the 'partistic' view, functional convergence of ecosystems (in its present meaning) started to fade away from written works until this broad topic was reduced to a mere discussion which species should be considered ecological equivalents (e.g. Begon et al., 2006). Even that was done, it seems to me, without any serious justification; the new generation of researchers simply shifted their focus towards other features of living nature. I consider this to be the most disappointing and costly mistake in the history of biology of the twentieth century.

Functional convergence of ecosystems suggests the existence of certain control or coordination at the level of ecosystem – otherwise it would be difficult to explain why different ecosystems all around the globe end up similar despite numerous casual factors, cataclysms, and differences in their mechanism of formation. At this point many of us are likely to get confused. Populations and communities seem to have no structures dedicated for control. It is worth noting, however, that control is not always centralised – it can also be diffused or passive (Novoselcev, 1978; Patten & Odum, 1981; see also Lekevičius 1986a, b), i.e. implemented through the interaction of system components that are equivalent in terms of control or, as it is often described, by way of self-organisation, spontaneously (Ashby, 1962). There are two indicators of self-organisation: the absence of an apparatus of control and the ability of the system to restore its organisation after minor disturbances by reverting to the same original state. Self-organisation is observed, for instance, in a protein molecule which spontaneously produces its secondary, tertiary, and quaternary structure. Self-organisation is also present in blastula and gastrula as well as in other stages of early embryonic development. Ecological succession can also be described as a process of self-organisation, albeit prolonged in time. Even the process of evolution itself is seen by some as a form of self-organisation mainly because it would be difficult to find any apparatus of control that sets relatively invariant paths of development.

Whenever we speak of control, it seems impossible without commands, restrictions, and permissions. These don't have to be written or oral, of course. Since the control apparatus is absent, there is just one possibility left: the command function must be performed by constraints that occur during the interaction of system components (Ashby, 1962; Salthe, 1985). Possibly the most important constraints are the negative feedbacks that keep the system stable. Individual components constrain and affect each other in such a way that only one or a few very similar self-organisation scenarios are chosen from among many hypothetical alternatives. According to Patten and Odum (1981),

> We hold that ecosystems are cybernetic (...) the cybernetic attributes emerge passively out of large and complex, decentralized system organization. The interplay of material cycles and energy flows, under informational control, generates self-organizing feedbacks with no discrete controller required.

It may be interesting to note that such diffused control was also observed in economics, in the wild capitalism prevailing at the times of the outstanding economist A. Smith (1723–1790) (Salthe, 1985). Eventually, the diffused control evolved into a partly-centralised system currently prevailing in most of the developed countries. In general, there are a striking number of similarities between organisation and control of ecosystems, on the one hand, and those of economics, on the other. Darwin has also noticed that, and it was A. Smith from whom he borrowed the idea that competition must not necessarily lead to the elimination of all weaker individuals but rather may lead towards a peaceful division of labour (niches) and specialisation (see the overview in Lekevičius, 2009).

So, do we have an answer to the question, why life has produced biological diversity? Not yet, for sure. However, we have clarified certain things.

References

Ashby, W. R. (1962). Principles of the self-organizing system. In H. von Foerster & G. W. Zopf Jr. (Eds.), *Principles of self-organization* (pp. 255–278). Pergamon Press.

Begon, M., Townsend, C. R., & Harper, J. L. (2006). *Ecology: From individuals to ecosystems* (4th ed.). Blackwell Publications.

Berryman, A. A., & Kindlmann, P. (2008). *Population systems: A general inroduction* (2nd ed.). Springer.

Cody, M. L. (1974). Optimization in ecology. *Science, 183*, 1156–1184.

Darlington, P. J., Jr. (1957). *Zoogeography: The geographical distribution of animals*. Wiley.

Duffy, J. E., Cardinale, B. J., France, K. E., McIntyre, P. B., Thébault, E., & Loreau, M. (2007). The functional role of biodiversity in ecosystems: Incorporating trophic complexity. *Ecology Letters, 10*, 522–538.

Ewel, J. J., & Bigelow, S. W. (1996). Plant life-forms and tropical ecosystem functioning. In G. Orians, R. Dirzo, & J. H. Cushman (Eds.), *Biodiversity and ecosystem processes in tropical forests* (Ecological studies) (Vol. 122, pp. 101–126). Springer.

Guimarães, P. R., Jr., Pires, M. M., Marquitti, F. M. D., & Raimundo, R. L. G. (2016). Ecology of mutualisms. In *Encyclopedia of life sciences* (pp. 1–9). Wiley.

Koestler, A. (1979). *Janus: A summing up*. Pan.

Kontrimavičius, V. L., & (Контримавичус В. Л.). (1982). Parasitism and the evolution of ecosystems. *Journal of General Biology (Moscow), 43*(3), 291–302. (in Russian).

Krebs, C. J. (2009). *Ecology: The experimental analysis of distribution and abundance*. Benjamin Cummings.

Lekevičius, E., & (Лекявичюс Э.). (1986a). Informational status of an ecosystem. In V. N. Maksimov (Ed.), *Ecological Forecast* (pp. 157–163). Moscow University Publisher. (in Russian).

Lekevičius, E., & (Лекявичюс Э.). (1986b). *Elements of a general adaptation theory*. Mokslas Publications. (in Russsian).

Lekevičius, E. (2000). *Only an ecosystem is alive: A somewhat nontraditional approach to the evolution of life*. Vilnius University. (in Lithuanian).

Lekevičius, E. (2002). *The origin of ecosystems by means of natural selection*. Institute of Ecology (Vilnius).

Lekevičius, E., & (Лекявичюс Э.). (2009). On some analogies between ecosystem evolution and development of economy: From A. Smith and C. Darwin to the latest ideas. In L. E. Grinin, A. V. Markov, & A. V. Korotaev (Eds.), *Evolution: Cosmic, biological, and social* (pp. 226–260). Librokom. (in Russian).

Loreau, M. (2010). *From populations to ecosystems: Theoretical foundations for a new ecological synthesis*. Princeton University Press.

Novoselcev, V. N., & (Новосельцев В. Н.). (1978). *Control theory and biosystems*. Nauka Publications. (in Russian).

Odum, E. P. (1953). *Fundamentals of ecology* (1st ed.). W. B. Saunders.

Odum, E. P. (1971). *Fundamentals of ecology* (3rd ed.). W. B. Saunders.

Odum, E. P., & Barret, G. W. (2005). *Fundamentals of ecology* (5th ed.). Thomson Brooks/Cole.

Pattee, H. H. (1968). Comments by H. In H. Pattee & I. C. H. Waddington (Eds.), *Towards a theoretical biology. Vol. 1. Prolegomena* (pp. 219–220). Edinburgh University Press.

Patten, B. C., & Odum, E. P. (1981). The cybernetic nature of ecosystems. *The American Naturalist, 118*, 886–895.

Pianka, E. R. (2000). *Evolutionary ecology* (6th ed.). Benjamin-Cummings.

Pimentel, D. (1968). Population regulation and genetic feedback. *Science, 159*, 1432–1437.

Polis, G. A., & Strong, D. R. (1996). Food web complexity and community dynamics. *American Naturalist, 147*, 813–846.

Reich, P. B., Walters, M. B., & Ellsworth, D. S. (1997). From tropics to tundra: Global convergence in plant functioning. *Proceedings of the National Academy of Sciences USA, 94*, 13730–13734.

Richards, P. W. (1957). *The tropical rain Forest. An ecological study*. Cambridge University Press.
Richardson, J. L. (1977). *Dimensions of ecology*. Williams and Wilkins Comp.
Robbins, J. (2004). Lessons from the wolf. *Scientific American., 290*(6), 44–49.
Salthe, S. N. (1985). *Evolving hierarchical systems: Their structure and representation*. Columbia University Press.
Schink, B. (2002). Synergistic interactions in the microbial world. *Antonie Van Leeuwenhoek, 81*, 257–261.
Shurin, J. B., Borer, E. T., Seabloom, E. W., Anderson, K., Blanchette, C. A., et al. (2002). A cross-ecosystem comparison of the strength of trophic cascades. *Ecology Letters, 5*, 785–791.
Smith, D. W., Peterson, R. O., & Houston, D. B. (2003). Yellowstone after wolves. *Bioscience, 54*, 330–340.
Taylor, L. L., Leake, J. R., Quirk, J., Hardy, K., Banwart, S. A., & Beerling, D. J. (2009). Biological weathering and the long-term carbon cycle: Integrating mycorrhizal evolution and function into the current paradigm. *Geobiology, 7*, 171–191.
Utida, S. (1957). Population fluctuation, an experimental and theoretical approach. *Cold Spring Harbor Symposia on Quantitative Biology, 22*, 139–151.
Valkiūnas, G. (2001). Practical importance of parasitic diseases: An outlook from the point of view of ecological parasitology. *Ekologija (Vilnius), 3*, 28–32.
Veldhuis, M. P., Berg, M. P., Loreau, M., & Olff, H. (2018). Ecological autocatalysis: A central principle in ecosystem organization? *Ecological Monographs, 88*, 304–319.
Whittaker, R. H. (1977). Evolution of species diversity in land communities. *Evolutionary Biology, 10*, 1–67.
Wilson, E. O. (1992). *The diversity of life*. Penguin Books.
Wilson, E. O. (2003). The encyclopedia of life. *Trends in Ecology and Evolution., 18*, 77–80.
Wilson, E. O., Eisner, T., Briggs, W. R., Dickerson, R. E., Metzenberg, R. L., et al. (1973). *Life on earth* (1st ed.). Sinauer Associates, Inc.

Chapter 4
Species Diversity as Trophic Specialisation

It is broadly accepted that every species or its population has not only specific habitat but also an ecological niche. This is one of the key concepts, however few other terms in ecology have such a vague and unsettled meaning. Sticking to the viewpoint of Odum (1971), I shall interpret the niche of a population not only as its needs, which are usually emphasised, but also as the totality of environmental factors in direct interaction with that population. Therefore, the description of an ecological niche should include the factors affected by the population as well as those that affect it, as well as the resources consumed, the other species that it competes with, and the species that exploit it as a resource. In other words, the niche of a given population is represented by its unique environmental conditions that the population is adapted to through evolution and coevolution.

An ecological niche comprises a biotic and an abiotic component (Fig. 4.1). The biotic niche of a population further consists of the following parts:

- Food consumed (not necessarily made of living organisms).
- Competitors.
- Mutualists (i.e. species that are in cooperative relations with the one under consideration).
- Predators and parasites exploiting the population.

Such definition of a niche may seem unusual to those readers who imagine a niche as a Hutchinsonian hypervolume. Nonetheless, this definition will be more appropriate for the purpose of this monograph, as most of us (hopefully) will soon be convinced. By the way, some other ecologists, including the famous Elton (e.g. Elton, 1927; Chesson, 2000), provide similar definitions of a niche.

One may speak about the niche of a single individual. It also must feed, compete for resources, may be exploited by others, and has to react to the abiotic factors of its environment. The individual niche is always more complex than that of the population as a unit because it also includes the interactions between its kin individuals: competition, complementation and/or cooperation.

The last chapter mentioned the division of functions of the local material cycle among functional kingdoms. One of the ideas was that trophic levels, guilds and individual species might be even more specialised, hence their trophic niches must be even more narrow. To illustrate this conclusion, let us take an example of decomposers. Detritus contains thousands of different organic substances. To digest each of them, specialised enzymes are necessary. Without them, digestions would be

E. Lekevičius, *Biodiversity*, https://doi.org/10.1007/978-3-031-11582-0_4

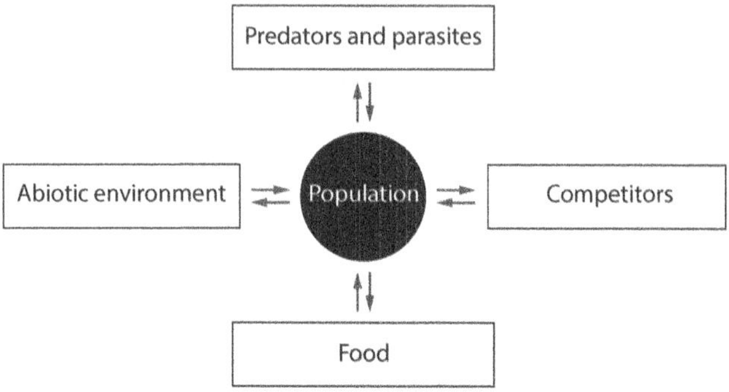

Fig. 4.1 A generalised model of ecological niche of a species (population) as its role in the ecosystem. The arrows pointing towards the population represent the environmental impacts it is subject to, whereas those pointing outwards represent the impact of the population on certain elements of its environment The arrows pointing upwards also represent material and energy flows. For the sake of simplicity, the picture excludes another possible interaction – mutualism, as well as time and living space which also are components describing the niche. (From Lekevičius 2013)

either impossible or painstakingly slow. Any decomposer species possesses only a limited set of such enzymes necessary for the digestion of these substances. Since all organic debris is sooner or later decomposed to the inorganic levels, one should assume that the totality of decomposers is equipped with a complete set of enzymes necessary for such a breakdown. Only in specific cases (for instance, during the formation of oil fields), when pH and/or temperature reach critical levels and/or the surrounding environment does not have enough materials serving as final electron acceptors in those specific conditions, the process of decomposition is practically halted. For instance, in summertime, the eutrophic zones of deep lakes produce significant quantities of planktonic detritus which slowly sinks to the bottom. At first, this detritus is partly decomposed by aerobic organisms, mostly bacteria. Then somewhat deeper, at the level of thermocline zone, detritus serves meals to facultative aerobic and anaerobic organisms. In addition to oxygen, they use weaker oxidants, such as nitrates. Even deeper, where oxygen concentrations become negligible, the baton is taken over by a guild which oxidises various substances by using the tetravalent manganese, an even weaker oxidant. As the depth increases further, decomposition of detritus involves other groups of organisms that use sulphates and trivalent iron. Finally, in the bottom sediments the process is continued by microorganisms that rely on carbon dioxide and acetate, the weakest oxidants of them all (Dodds, 2002).

Wilson (1992, pp. 195–198) raises the question why there are millions of arthropod species in wild nature, and immediately comes up with the answer: most likely, wild nature provides millions of different ecological niches that are suitable to them. This implies that a niche of every single species is almost unique. Can such enormous quantity of larger and smaller niches be real? Yes, it can, according to Wilson. He goes on by mentioning a well-known fact that the numbers of small organisms (species) by far exceed those of big ones, and this regularity seems to apply to both vascular plants and algae, and to animals. He calls this regularity the pyramid of diversity. It can be explained as follows: to small organisms even a palm-sized area may become an entire ecosystem with a large number of different niches that can sustain thousands of species. Imagine, says Wilson, a 50 mm long tropical beetle crawling on a trunk of a large tree and grazing on lichens and fungi. It does not give a damn about what is happening under its feet. Meanwhile, other tiny beetles less than 1 mm long are scuttling around down there. They live in tiny hollows, burrows, and under the bark and might travel no more than a few meters during their lifetime. Different nooks and holes have different lighting, humidity, and temperatures, different algae and fungi plantations serving food to these little beetles, and specific combination of all these factors meets the needs of a certain beetle species. Most of these species would not even be known to entomologists. On the other hand, the bark has many even smaller cavities that are beyond reach even to these beetles. They are the realm of the tiniest insects and oribatid mites, while dozens of bacterial species thrive under their feet. All these organisms manage to find their trophic and abiotic niche there.

Yet another example (Wilson, 1992, p. 198): the plumage of any bird serves home to at least a few or a dozen of feather mite species, and different bird species would host different sets of these species. Some feed upon oily secretions and chips splitting from the cells, and they are specialised according to the type of feather and its location while some others are parasites. These mites might spend their entire life within a territory of a few square centimetres, and thus they have had ideal conditions for such differentiation.

Ricklefs (1979, p. 419) has come to a similar conclusion:

> To large species, the habitat is relatively more homogeneous and presents less opportunity for adaptive radiation than it presents to small species. Following this argument, it is no surprise that there are many more kinds of insects than birds and mammals.

I think these authors are right: specialisation in nature is much more delicate than we originally thought. The niches of coexisting organisms, and thus their needs and impact on their environment differ both in qualitative and in quantitative terms. We may perceive the niches of the same guild species that share the same territory as identical and thus find it difficult to explain their coexistence, but only as long as we restrict ourselves to a superficial description without going into quantification. I hope most ecologists would support this position. However, only proper consideration of this detailed quantitative aspect makes it possible to imagine theoretically almost unlimited number of possible different niches on Earth as combinations of environmental parameters that may each meet the demands of some present or future organisms.

By the way, to support these conclusions, Wilson (1992) refers to, among others, the studies by Erwin (1982) who, together with his colleagues, collected tropical forest insects of Central and South America with the help of a 'bug bomb' – a device that would send a cloud of insecticide spray towards forest canopy. All that remained to do was to collect and classify all the insects and other arthropods fallen from the trees. Erwin chose *Luehea seemannii* – a tree species close to the linden tree. He discovered 1200 insect species on them. Among these, 163 species could be found exclusively on these trees, most also feeding on them, although some were carnivorous or fed on fungi.

Nevertheless, I would like to supplement the words of E. O. Wilson. One should not think that the number of species in the biosphere equals the number of different niches. This possibility is denied by the phenomenon of functional convergence of ecosystems mentioned in the previous chapter. World's ecological communities are organised according to the same template. Creativity of nature has been quite limited in this case. Especially high number of similarities can be observed when comparing communities of the same biome. They resemble those residential apartment blocks, which are inhabited by various people who differ both among themselves and from the inhabitants of the neighbouring buildings, while the buildings themselves and the apartments within them are standardised and designed by the same architect. Thus, we should get accustomed to the idea that each species of our biosphere (or most of them) might have at least a few ecological equivalents outside their usual habitats, and in many cases the numbers of these equivalents may reach hundreds. Apparently, geographical isolation was quite sufficient for the occurrence of equivalent species with different histories of evolution in similar niches of the communities physically isolated from one another. On the other hand, isolation was clearly not enough to prevent migration from bringing those equivalents together sooner or later, which would result in the displacement of the weaker competitors, the scenario foreseen by Darwin himself (1872).

Let me add a clarification. Ecologists usually perceive equivalents as the species occupying similar but not necessarily identical niches. A classic example could be marsupials of Australia and their placental equivalents elsewhere, although it is well known that foraging niches of these equivalents are not completely identical. Nonetheless, there is multiple evidence that these animals have difficulties to coexist when placed together in the same locations, which suggests great similarity of their niches. I have no doubt that each of my readers, especially those who are experts in biodiversity, could come up with comparable examples, since these facts have become quite trivial by now. On the other hand, I could not find any references to studies that would attempt to provide at least rough estimates of the number of possible equivalents, say, among plants, herbivores, or large carnivores. This issue does not seem to bother ecologists or evolutionary scientists; in fact, any reference is so difficult to find as if the problem does not exist. One of the reasons may be the lack of tradition to describe the niches of species following a more detailed version of the scheme presented in Fig. 4.1. There might be other reasons behind it as well, such as objective difficulties to obtain such data. The lack of data may also explain why forecasting the consequences of natural or artificial introduction of plants and animals is such a challenge.

There is no doubt that these consequences might be quite dramatic. Exchange of species between the flora and fauna of different geographical regions results in the displacement of many weaker equivalents leading to the loss of biodiversity and biotic homogenisation on the global scale (Sax & Gaines, 2003). A species that immigrates to a new territory and finds a suitable niche there might not necessarily be less adapted to that niche than the resident equivalent species. If peaceful sharing of the niche turns out to be unlikely, displacement is quite possible. Success of invasive species in the modern biosphere is rather widespread phenomenon which can be explained as follows: successive invasive species have undergone longer evolutionary development, their populations carry higher genotype diversity, or they have been exposed to a higher number of competitors in their evolution (Fridley & Sax, 2014). By the way, Darwin also followed a similar line of reasoning (1872).

But let us return to the main topic and discuss individual functional groups. Vegetal secondary compounds are considered to be one of the factors supporting the diversity of herbivores, both in tropical areas and beyond (Loxdale et al., 2011; Loxdale & Harvey, 2016). These are various substances produced in defence against herbivores (alkaloids, terpenes, flavonoids, tannins, lignins, etc.). We have identified tens of thousands of these substances (Wink, 2010), however, there is no doubt that their actual numbers could reach hundreds of thousands. Herbivores normally are only able to feed on a limited number of plant species whose secondary compounds are non-toxic to them or they are capable to break them down or neutralise in some other way. Thus, these herbivores avoid competition with others who are grazing on different plant species. It is obvious that if plants did not produce such diversity of secondary compounds each herbivore species would have access to a larger number of plants, which would lead to the convergence of foraging niches of many herbivores, intensified competition between them, and, most likely, decrease of their diversity. On the contrary, when a plant species starts producing a new secondary compound due to some mutation, diversity only rises, since this new genotype may repel herbivore attacks, proliferate, and eventually become a new species or even genus. It is very likely, however, that some time later certain herbivores will likewise acquire an ability to neutralise the new compound and gain a competitive advantage leading to possible new species. The original species in other parts of their range will remain unchanged, so the overall number of species will increase. Such scenario is presented by the theory of co-evolution (Ehrlich & Raven, 1964; Futuyma & Agrawal, 2009).

Another group of herbivores, the pollinating insects and birds, are so closely linked with their symbiont plants that disappearance of one group would lead to the extinction of the other. In this case, it is difficult to tell which of the two occurred first, the blooming plants or their pollinators, but beyond doubt these partners have been evolving together shoulder-to-shoulder. These ties are especially strong among solitary bees – there are over 18 000 known species that pollinate even larger number of plant species, although in most cases a single plant species has its own specialised bee pollinator.

Ecologists try to explain the enormous diversity of herbivorous insects which is not limited to the tropical areas by their specialisation to feed on different plant species, their different genotype varieties, different organs or tissues of the same plant, and at different locations and times. For instance, Crutsinger and his team (Crutsinger et al., 2006) have identified 21 different genotypes within a wild population of *Solidago altissima* plant by using the amplified fragment length method. Then they designated 63 experimental plots with different levels of polymorphism and observed the numbers and species of arthropods visiting these plots during the vegetation period. It turned out that the diversity of both herbivorous and carnivorous arthropods was increasing with the increase of genotype diversity of *S. altissima* in the corresponding fields. The authors concluded that the herbivores were highly specialised: some species preferred to exploit certain genotypes of the host plant, while the others opted for different ones. It is quite possible that the results of these researchers will help us one day explain why the diversity of insects is much higher in tropical areas as compared to temperate ones. Some scholars hold an opinion that trophic niches of tropical herbivores are extremely narrow (Strong et al., 1984; Condon et al., 2014; Forister et al., 2015), although generalists can also be found among them.

So, some 300,000 plant species support, presumably, about a million insect species. Erwin (1982) is convinced that there must be about 31 million species of arthropods alone on our planet, and most of them are insects. However, some broader studies carried out afterwards (e.g. Novotny et al., 2006) have demonstrated that tropical herbivorous insects are far less specialised than Erwin assumed, and their β-diversity was lower than previously thought (fewer endemic species). After the corresponding adjustments, the estimate of 31 million arthropod species was truncated to 'mere' 4–6 million. Mora with collaborators (2011) have calculated that Earth should host about 8.7 million eukaryote species, most of them being arthropods. Of course, this is not the final estimate.

It seems natural to assume that the diversity of predators, parasites, and parasitoids is correspondingly supported by the diversity of their prey or host organisms. Nonetheless, the numbers of predators in nature are a couple of orders of magnitude smaller than those of parasites and parasitoids, which is understood, because the latter two groups are much more specialised in terms of food than predators. There are no organisms completely free of parasites, and most of them host dozens of them. Some parasitologists (Windsor, 1998) consider that about half of all species on Earth lead a parasitic lifestyle. Calculations suggest that humans, as a species, is exploited by more than 1500 different parasites, and new species are being discovered annually. Most human parasites are bacteria, viruses, fungi, protists, and helminths. Specialised ectoparasites exploit different host species, however, different parasites can also be found on different body areas of a single individual. A good example of the latter case could be the head louse, pubic louse, and body louse, or the already-mentioned feather mites. The same stands true for endoparasites: different species are often specialised to live in different host species or even different organs and tissues of the same host individual. Even speaking about parasites that are referred to as generalists (because of their relatively broad trophic niche), we must admit that their diversity depends on the diversity of actual and potential hosts. Once these get extinct, their parasites will meet the same fate. We should also bear in mind the fact that phylogenetically and physiologically closely related species can be, and often are, successfully infested by other genotypes of the same pathogenic species. A god example of that are pathogenic strains of *Escherichia coli*, each of whom are specialised to exploit a specific species of endothermic animals, including humans and domestic animals. A similar polymorphism is observed in the populations of another generalist, the *Rabies* virus. This helps the generalist parasites to maintain their intra-species (genotype) diversity. For more on parasite and parasitoid diversity and their specialisation, see Meeus et al. (1998), Woolhouse et al. (2001, 2005), and Loxdale and Harvey (2016).

Of course, these conclusions should not be interpreted as suggesting that the number of food sources in nature coincides with the number of species that exploit them. Monophagous organisms are rare in nature. The available data indicate that a single species is usually exploited by at least several other organisms, while these consumers themselves usually feed on more than one species (Polis & Strong, 1996). The only regular pattern, if any, is that herbivorous species are more numerous than plant species, and the number of parasite species is larger than that of the host organisms. Predators seem to be the only exception, being less diverse than their prey, because their trophic niches are relatively broad.

All that suggests that differentiation of trophic niches, or resource partitioning in other words, is one of the key prerequisites for the peaceful coexistence of species within communities. According to Chesson (2000), such specialisation clearly leads towards the intra-specific relationships being more antagonistic than the inter-specific ones, and this makes coexistence possible. See also Chesson and Kuang (2008).

References

Chesson, P. (2000). Mechanisms of maintenance of species diversity. *Annual Review of Ecology and Systematics, 31*, 343–366.

Chesson, P., & Kuang, J. J. (2008). The interaction between predation and competition. *Nature, 456*, 235–238.

Condon, M. A., Scheffer, S. J., Lewis, M. L., Wharton, R., Adams, D. C., & Forbes, A. A. (2014). Lethal interactions between parasites and prey increase niche diversity in a tropical community. *Science, 343*, 1240–1244.

Crutsinger, G. M., Collins, M. D., Fordyce, J. A., Gompert, Z., Nice, C. C., & Sanders, N. J. (2006). Plant genotypic diversity predicts community structure and governs an ecosystem process. *Science, 313*, 966–968.

Darwin, C. (1872). *The origin of species by means of natural selection*, 6th ed. John Murray (Printed in 1998 by Cox & Wyman. Reading).

Dodds, W. K. (2002). *Freshwater ecology: Concepts and environmental applications*. Academic Press.

Ehrlich, P. R., & Raven, P. H. (1964). Butterflies and plants: A study of coevolution. *Evolution, 18*, 586–608.

Elton, C. (1927). *Animal Ecology*. Sidgwick and Jackson.

Erwin, T. (1982). Tropical forests: Their richness in Coleoptera and other arthropod species. Coleopterists'. *Bulletin, 36*, 74–75.

Forister, M. L., Novotny, V., Panorska, A. K., Baje, L., Basset, Y., et al. (2015). The global distribution of diet breadth in insect herbivores. *Proceedings of the National Academy of Sciences USA, 112*, 442–447.

Fridley, J. D., & Sax, D. F. (2014). The imbalance of nature: Revisiting a Darwinian framework for invasion biology. *Global Ecology and Biogeography, 23*, 1157–1166.

Futuyma, D. J., & Agrawal, A. A. (2009). Macroevolution and the biological diversity of plants and herbivores. *Proceedings of the National Academy of Sciences USA, 106*, 18054–18061.

Lekevičius, E. (2013). *Ecology: From individuals to the biosphere*. Vilnius University Publications. (in Lithuanian).
Loxdale, H. D., & Harvey, J. A. (2016). The 'generalism' debate: Misinterpreting the term in the empirical literature focusing on dietary breadth in insects. *Biological Journal of the Linnean Society, 119*, 265–282.
Loxdale, H. D., Lushai, G., & Harvey, J. A. (2011). The evolutionary improbability of 'generalism' in nature, with special reference to insects. *Biological Journal of the Linnean Society, 103*, 1–18.
Meeus de, T., Michalakis, Y., & Renaud, F. (1998). Santa Rosalia revisited: Or why are there so many kinds of parasites in the garden of earthly delights. *Parasitology Today, 14*, 10–13.
Mora, C., Tittensor, D. P., Adl, S., Simpson, A. G. B., & Worm, B. (2011). How many species are there on earth and in the ocean? *PLoS Biology, 9*(8), e1001127. https://doi.org/10.1371/journal.pbio.1001127
Novotny, V., Drozd, P., Miller, S. E., Kulfan, M., Janda, M., et al. (2006). Why are there so many species of herbivorous insects in tropical rainforests? *Science, 313*, 1115–1118.
Odum, E. P. (1971). *Fundamentals of ecology* (3rd ed.). W. B. Saunders.
Polis, G. A., & Strong, D. R. (1996). Food web complexity and community dynamics. *American Naturalist, 147*, 813–846.
Ricklefs, R. E. (1979). Ecology. (2d ed.). Chiron Press.
Sax, D. F., & Gaines, S. D. (2003). Species diversity: From global decreases to local increases. *Trends in Ecology and Evolution, 18*, 561–566.
Strong, D. R., Lawton, J. H., & Southwood, R. (1984). *Insects on plants*. Harvard University Press.
Wilson, E. O. (1992). *The diversity of life*. Penguin Books.
Windsor, D. A. (1998). Most of the species on earth are parasites. *International Journal of Parasitology, 28*, 1939–1941.
Wink, M. (2010). Introduction: Biochemistry, physiology and ecological functions of secondary metabolites. *Annual Plant Reviews, 40*, 1–19.
Woolhouse, M. E., Taylor, L. H., & Haydon, D. T. (2001). Population biology of multihost pathogens. *Science, 292*, 1109–1112.
Woolhouse, M. E. J., Haydon, D. T., & Antia, R. (2005). Emerging pathogens: The epidemiology and evolution of species jumps. *Trends in Ecology and Evolution., 20*, 238–244.

Chapter 5
Producer Diversity as Reflection of Variation in the Abiotic Environment

The biggest challenges arise with the attempt to understand the relations established within a trophic level or a guild. For a long time, the prevailing opinion was that at these levels we observe a clear prevalence of competition. After all, in this case trophic niches are very similar, and some may even overlap. Admittedly, an overlap is usually never complete, and the non-overlapping boundaries (or the mismatch of the optima) determine the extent to which species complement each other (Whittaker, 1975). Of course, this complementarity does not imply any obligatory nature: the interacting species could perfectly thrive without one another. Nonetheless, about two decades ago some scientists started to voice the opinions that even within trophic levels and guilds one may find almost obligatory dependence (see Ewel & Bigelow, 1996; Callaway, 2007; Brooker et al., 2008; McIntire & Fajardo, 2014). For instance, large woody plants obscure the sunlight thus creating favourable conditions for the growth and proliferation of shade-tolerant plants. They also create microclimate that shelters those plants from the wind and cold and also prevents the desiccation of soil. Moreover, a grown-up plant itself usually becomes a set of vastly different habitats suitable for algae, fungi, epiphytes, and vines. Many of these tenants are unilaterally dependent on their host, they would not survive without it. Even the relationship between competing species may not be exclusively adverse: a species A that supresses a species B might help a species C to settle in and to coexist, if the latter does not compete much with the first one but is strongly suppressed by the species B (Laird & Schamp, 2006).

The authors consider similar cases as examples of direct and indirect facilitation. Facilitation is understood in this context as an interaction wherein a species alters its environment in such a way that it becomes favourable for the arrival and survival of other species. Moreover, it may become a habitat suitable for other species. Such views are sometimes called the theory of ecological facilitation. The term is taken from a hypothesis sharing the same name aimed at the explanation of the mechanisms of primary succession (Connell & Slatyer, 1977). McIntire and Fajardo (2014) conclude with the following outstanding summary of their views:

> While biodiversity is, in part, created by species exploiting many niches, many niches are available to exploit only because species create them (...) facilitation is actually a major driver of biodiversity across ecosystems.

From year to year, the numbers of articles on facilitation grow exponentially. I may be wrong, but I have an impression that some revolutionary changes are currently

E. Lekevičius, *Biodiversity*, https://doi.org/10.1007/978-3-031-11582-0_5

taking place in the theory of ecology, and I would not be surprised at all to see the long prevalent individualistic (or competition) paradigm being replaced by a new approach which states that an opposite force, that of biotic attraction, is also present in the living nature, and that it is no less important in shaping biodiversity. However, I should remind the reader that even Clements (1916), one of the most outstanding ecologists of all times, wrote about functional interdependence and division of labour at the level of community. Admittedly, his original views (the concept of a superorganism) have been eventually replaced by the individualistic doctrine. Somewhat later they were resurrected in slightly modified form in the ecosystem concept, just to lose most of their positions again as the time passed by. Is it the ripe time to bring back to memory F. Clements and E. P. Odum again? I do not think that such swinging between two extremes demonstrates the solidity of ecological science. And yet I cannot but welcome the renaissance of the concept of facilitation.

A single drop of lake water may contain a few dozens of algae and cyanobacteria species. For many years, this fact has been pushing ecologists to desperation. They presumed that the needs of all phytoplankton species were identical or at least very similar: they all need light and the same nutrients. Another comparable case is zooplankton, especially the filter-feeding cladocerans. They all manage to coexist peacefully, although they all appear to be sharing the same food: algae and bacteria. In both cases the fact of lasting coexistence seems to go against the principle of competitive exclusion discovered by G. F. Gause back in 1934 as he performed experiments on *Paramecium* protozoans. Hutchinson (1961) named this mismatch between theory and reality *the paradox of the plankton*.

Hutchinson also suggested his own solution to this paradox. In his opinion, species diversity of the plankton is sustained by the variation of environmental conditions with regard to time and space. Although the first impression might be that the niches of planktonic organisms are identical, in reality they are similar but not exactly the same. There are differences not only in feeding needs but also in the optimum temperatures, pH values, and other factors. These differences make coexistence possible. In other words, planktonic organisms divide the apparently uniform adaptive zone into much smaller components representing their niches.

In fact, hydrobiologists have been long aware about the annual seasonal succession which takes place in the lakes of temperate climatic zone. Diatomic algae usually prevail early in the vegetation season, later they are replaced by green algae, while during the peak of the heat cyanobacteria account for the major part of the biomass. Laboratory tests revealed that these groups of organisms have different light- and thermopreferenda (optimal light and temperature ranges). As these environmental factors show variations in the course of a year, different species gain and loose advantage, and this might be the likely reason of their coexistence.

Experiments were needed for more accurate conclusions. One of the most successful ones was performed by Tilman (1977). His experiment was very similar to that of G. F. Gause. The only difference was that this time the researcher was more interested in the reasons of species coexisting without eliminating each other, rather than in the reasons of successful elimination. Tilman would place two species of diatoms, *Asterionella formosa* and *Cyclotella meneghiniana*, into media with different silicon and phosphorus concentrations. Where silicon was easily available, *Asterionella* sooner or later would eliminate *C. meneghiniana*. With the shortage of silicon, the result was exactly the opposite. And only in cases when the concentrations of both nutrients were intermediary both species were able to coexist without problems. It seems that the feeding niches of these two species are different: one of them (*A. formosa*) needs large quantities of silicon and can tolerate lower concentrations of phosphorus, whereas its competitor has much higher demand for phosphorus. Under natural conditions, concentrations of these two nutrients vary in time and space, therefore none of the two species gains a permanent advantage.

What would happen then if this environmental variation is eliminated? Coexistence should become impossible. This hypothesis can be experimentally tested. Indeed, such tests have been carried out. Sommer's work (1984, 1996) stands out for its elegance. He performed his experiments in a chemostat designed for phytoplankton cultivation.

The algae (some 30 species) were taken from Lake Constance in Germany. The temperature within the chemostat was set at 18 °C. The control group had constant quantities of phosphorus and other nutrients for the entire duration of the experiment. The experimental group had a phosphorus-deprived medium where phosphorus was added in weekly portions, and then its concentration would steadily decline till the next replenishment. The experiments lasted for a few weeks and at the end the number of surviving species was assessed in comparison with the initial number of species. The research revealed that species diversity was gradually declining in both control and experimental groups, yet the extent of this decline was different. While the control group would often contain just a single species by the end of the experiment out of the initial 30, the experimental group would sustain 5–7 of them. The author concludes that even higher species diversity could be preserved in the chemostat, possibly even comparable to the wild communities, if the variable parameters were not limited to phosphorus but would include other abiotic and biotic factors as observed in natural environment.

These results and conclusions appeared even more solid after the experiment performed by Flöder and his co-authors (2002). They took a phytoplankton sample from Lake Biwa in Japan and divided it into several sub-sets that were then incubated under different lighting regimes:

(a) Under permanent bright light.
(b) Under permanent light which was 5 times less intense.
(c) Under variable lighting, shifting from dim to bright light and then backwards, with the periods of 1, 3, 6, and 12 days.

The nourishing medium was sufficiently rich, and its composition was kept stable throughout. The experiment lasted for 49 days. Just as expected, the numbers of phytoplankton species were declining in all the 6 experimental sets, however, by different degree. The most species survived in the set with the period of light fluctuation set to 6 days. Both shorter and longer periods of lighting reduced the species diversity of the guild even more. The authors concluded that phytoplankton species have different levels of need for light and minimum lighting, necessary for their survival and reproduction. The period of six days, roughly comparable to 3 generations, is quite sufficient to allow the plankton species that acquire temporary advantage to increase their populations and to start dominating; however, this advantage does not last long enough to allow a complete elimination of other species that are prevalent in different phase of the cycle. As the period is prolonged to 12 days, the displacement does happen. When the conditions fluctuate too fast (every 1 or 3 days), ecological succession cannot proceed, although the species with low phenotypic plasticity get eliminated. According to the authors, the 6-day periodicity is the closest to the fluctuations also observed in natural environment.

A much bigger challenge to ecologists was the attempt to identify the forces upholding the diversity of land producers, especially in tropical latitudes. The very extent of this diversity is striking: a single hectare of Amazon forest contains over 300 tree species, and a single square kilometre may have over 2000 of them. A tropical forest of 100,000 km^2 may include dozens of thousands of plant species! Clearly, β-diversity in this biome is also quite high.

At the first sight, the niches of all coexisting plants appear to be very similar – they all need light, the same nutrients and water, just like in the case of the plankton. It seems quite logical to expect that the number of ecological equivalents among plants should be much higher than among other organisms. Thousands of allopatric equivalents per one niche? But why not? I think this may be a very real possibility. And we should have little doubt that elimination of geographical barriers would result in biotic homogenisation and dramatic global decline of biodiversity, as mentioned already.

A much trickier question is whether sympatric equivalents can coexist within a single local phytocenosis, although there are some authors (e.g. Hubbell, 2001) who deem it possible, too. If so, how can this fact go along with the Gause's principle? Ewel and Bigelow (1996) made a justified statement when discussing the guilds, or life-forms, of tropical plants:

'If, among higher plant species, there is substantial redundancy in nature, then loss of a minor player or substitution of one species for another is likely to have immeasurably small consequences for mesoscale processes such as carbon and oxygen exchange, soil erosion, and water and nutrient budgets. But this is not likely to be the case with loss of life-forms, for if there are significant functional differences among plants, surely these will manifest themselves most dramatically among species of strikingly different architecture. It is the assemblage of life-forms that give forests their characteristic structure, and structure, in turn, dictates whole-system functioning'.

The authors mean that functional differences between the guilds of tropical plants (trees, herbs, epiphytes, climbers, treelets, shrubs, giant herbs, vines, epiphytes, forbs, grasses, and succulents) are obvious, but the niches of species within the same guild clearly overlap, therefore peaceful coexistence of species within a guild is much more difficult to explain. One could say without any exaggeration that this problem has become a major headache to ecologists, an enigma still awaiting the ultimate resolution.

Before we focus on the tropics, let us take an easier example of a more homogenous system of the temperate plant communities. The botanists who study them (Ellenberg, 1974; Whittaker, 1975) have no doubt that the niches of plants coexisting within a limited area are not identical, these plants have different needs and are subject to different environmental factors. Some are light-demanding species, while others might be shade-tolerant or stand somewhere in-between; some species demand more moisture, others less; some tolerate heat easily, while others are more sensitive to it; some prefer acidic soils and others alkaline. Different species absorb their nutrients from different soil horizons; some of them have high demands for a particular nutrient in their soil, while others may be less dependent on it or can tolerate its absence, because they are capable to get it from the air with the help of bacteria, and so on, and so forth. To put it short, this climatic zone has so diverse abiotic conditions that each plant is able to find its appropriate niche. Even when we observe several herbaceous plants smoothly coexisting within a plot of a mere square metre, we always seem to know the explanation: 'they must be absorbing nutrients from different soil horizons and thus peacefully share the niches'.

There is another possible answer: the fact of these plants coexisting might indicate that the soil composition is not homogenous across the entire plot. Indeed, it is well known that even relatively small differences in soil acidity, humus content, or levels of the ground water might be the reason we find quite diverse plant associations within a fairly small forest area. This differentiation persists regardless of a regular exchange of seeds between the neighbouring forest plots. One might assume that the special edaphic diversity, sometimes referred to as the patchiness of soil, is one of the preconditions for the emergence and preservation of species diversity. Admittedly, this diversity, especially of tree species, may be rather limited in some areas, for instance, in Lithuania, and their edaphic niches are relatively broad. E.g. *Pinus sylvestris* can grow in soils with pH ranging from 3.1 to 4.8. This interval overlaps with the limits suitable to some other tree species growing in the same area.

Could patchiness of soil be the explanation behind the diversity of tropical vegetation? Jordan (1985) attempted to answer this question. He chose the Amazon forests for his research. He concluded that even small territories host more than one plant association, and their existence is most likely determined by the soil qualities and ground water levels. Higher areas with quality soils usually support mixed forest rich in plant species. Lower areas with deeper clay horizons have less diverse forest predominated by the *yévaro* trees (*Eperua purpurea*). The floodplain along a stream gives rise to another association, the *igapó* forest. Poor soils with relatively high ground water level but beyond the flood zone support the *caatinga* association. Even higher, where sandy soils are not capable to withhold the rainwater and soil humidity is low, Jordan found a shrubbery called *bana* by the locals.

Could these data serve as the key to the mystery chest called tropical biodiversity? Hardly so, since a similar soil diversity can also be found in the temperate climate zone (see Table 5.1), although the number of species is ten times lower there.

Tropical rainforests have a more pronounced vertical gradient of light intensity than temperate forests. Therefore, only highly shade-tolerant plants can grow in the underwood and the lower layer. Could this gradient offer hundreds of different lighting niches to the coexisting plants, so that each species would have one for itself? Very unlikely, especially considering the obvious fact that the lighting needs of a particular species are nor carved in stone for their lifetime and they often change as the plant develops. This is especially true for the understory, the woody plants with broad intervals of light-tolerance that must overlap to some extent, although they hardly overlap completely.

Plants of the equatorial rainforests also enjoy abiotic environment that is exceptionally stable in time (see Table 5.1): the average annual temperature is 24–28 °C; there is no seasonal variation and the annual temperature amplitude is 2–3 °C, which is often less than the circadian (daily) amplitude (up to 5 °C); the amount of sunlight remains almost unchanged throughout the year. The temperature niches of local organisms seem to be identical or to significantly overlap. Different tropical species have no necessity for wide range of temperature optima so common in the temperate climate zone.

Some researchers (Kelly & Bowler, 2002) have expressed the idea that α diversity of plants could be supported by the fluctuation of environmental conditions in time in a similar way as it happens in plankton. According to the authors, the niches of coexisting plants appear identical only at the first sight; therefore, as the environmental conditions fluctuate from year to year, different species enjoy advantage now and again but none of them is able to gain a long-term superiority and eliminate the others. This is the 'storage effect' hypothesis. So far, it has failed to achieve broad support, even if, at first sight, it seems quite reasonable. For instance, in Lithuania, the fluctuation of the mean annual air temperature for April in the period of 2013–2018 was in the range of 5 °C, while that for March was even higher, in the range of 10 °C (http://www.hkk.gf.vu.lt/vu_ms/duomenu-suvestine/). It is well known that in spring, when spores and seeds germinate, and sprouts take root, they become especially vulnerable to any climatic deviations from 'the norm'. However, these interannual fluctuations seem to be able to cause only differential changes in numbers of individual plants within communities and they are unlikely to maintain

Table 5.1 Characteristics of niches typical for temperate forests and equatorial rainforests

Ecological sub-niches	Temperate zone	Equatorial zone
Seasonal and day-to-day variation of humidity (precipitation)	Variation *high*	Variation *low*
Seasonal variation of temperature	Variation *high*	Variation *low*
Seasonal variation of light intensity	Variation *high*	Variation *low*
Vertical gradient of light intensity between the layers of vegetation	*Conspicuous*	*Even more conspicuous (wider range)*
Vertical gradient of nutrient composition and concentrations in soil	*Conspicuous*, reaching over one metre deep	*Less conspicuous*, nutrients almost absent already at the depth of 20 cm
Diversity of soil composition, structure and pH values (soil patchiness)	*High* in both zones	

the number of species therein. To achieve a positive impact of this fluctuation on species diversity, the oscillation period should be decades or even centuries, rather than a single year (as in the case of the plankton paradox, see the discussion in Lekevičius & Loreau, 2012). In any case, the storage effect hypothesis cannot explain the extremely high diversity of tropical plants (interannual climatic fluctuations are negligible in tropics). On the other hand, the mentioned interannual climatic fluctuations appear to be sufficient to support genetic diversity within plant populations (see Chap. 10).

At least a few dozens of hypotheses have been proposed to explain the stable coexistence of multiple tropical plants, but no one has come up with a definite answer yet. Comparison of niches (Table 5.1) does not support the idea that variation and diversity of environmental conditions are higher in tropics than in the temperate zone. The diversity of tropical plants is much more likely to be supported by the extremely intense sunlight, high levels of primary production as well as abundant and regular year-round precipitation, all typical for that region, while soil and topographic heterogenicity are less important (Kreft & Jetz, 2007). This may only imply (and most ecologists seem to be moving towards this idea) that the niches of tropical plants are much narrower than those of the temperate zone vegetation (Tilman, 1986; Tilman & Downing, 1994; Sterck et al., 2011). Besides, the niches of coexisting tropical (as well as temperate) plants must differ, which might enable these species to grow together. This is supported by numerous studies performed by such well-known authors as R. E. Ricklefs, F. A. Bazazz, J. S. Denslow, T. Kohyama, P. M. Vitousek, G. H. Orians, D. B. Clark & D. A. Clark, etc.

Among the fieldwork aimed at understanding the reasons of this enormous variety of trees in tropical rainforests, I would point out an especially detailed research carried out in the island of Borneo (Davies et al., 1998). It focused on 11 species of the *Macaranga* genus (the *Euphorbiaceae* family) coexisting in one relatively small territory. Over 4000 individual plants were studied, along with the natural resources in their locations. It turned out that these species varied significantly in terms of their needs: they included light-demanding species that eventually grow into tall trees, smaller shade-tolerant species, and species with intermediate needs. Some preferred sandy while others clay soils. Species with the very similar needs for light differed by their soil preferences and, on the contrary, species growing together in a sandy (or clay) soil would usually have different needs for light intensity. Their seeds also varied according to their ability to germinate and survive in forest clearings resulting from some localised destruction (e.g. collapse of a large tree). These data correlate well with the facts previously collected by other authors (see references in Davies et al., 1998) on factors that might support coexistence of tropical trees. In recent years, even more detailed and extensive studies of a similar type (see e.g. Baldeck et al., 2013, and references therein) led to the conclusion that, in addition to the previously mentioned factors, coexistence of tropical trees is also determined by the diversity of soil chemical composition (quantities of micro and macro elements) and topography (elevation, surface slope, and its direction).

Simple calculations show that even a few gradients of environmental conditions determining vital functions of plants are enough to support a relatively high species diversity, provided that the niches of these species are sufficiently narrow. Let us assume that there are three types of environmental diversity that determine α-diversity of tree species in the equatorial zone (the vertical gradient of temperature and light, the vertical gradient of nutrients composition and quantities in the soil and the topographic diversity). If each type of diversity creates a gradient that allows 10 plant species to coexist without disturbance, the total theoretically possible number of coexisting species is $10^3 = 1000$. Thus, if there are no limits to the fragmentation of niches or degree of specialisation, in theory, tropical α-diversity could be even higher than that actually observed.

Of course, this picture is a very schematic vision, albeit quite realistic. I think that Tilman's (2004) 'stochastic niche theory' offers a much more refined – though more difficult to grasp – explanation why plants can peacefully coexist. Based on his own multi-annual field experiments, Tilman draws the following conclusions. Firstly, the more plant species are found in a particular area, the less available resources remain, and the lower the possibility for new species to settle in that area. When no free resources are left, new arrivals can no longer establish themselves. Secondly, a new species can establish itself even in a species-rich community, provided that its demands for resources differ somewhat from those of the resident species. This conclusion should be interpreted as follows: the nutritional niches of all coexisting plants must be similar in the sense that they all need the same nutrients and light; however, Tilman places emphasis on quantitative rather than qualitative differences in needs. In nature, we usually observe not a complete absence or presence of that or other resource, but gradients of nutrient availability that further vary in space and time, which does not allow any single species to gain advantage.

Even earlier Tilman (1994, 1997) presented a few more conclusions derived from his own field experiments carried out with his co-workers. These experiments show that some plant communities, such as the wild prairies (which have become really sparse in America), are not saturated by species because the researchers could significantly increase the number of species in local communities by introducing seeds of other species taken from the neighbouring communities. It also turned out that these new species not only manage to grow and produce seeds but also produce no significant negative impact on the biomass of the resident species. Tilman concludes that this indicates the availability of 'empty' slots (or unoccupied niches, as I would call them) in the community. Why then the plants were not able to achieve the same result, a successful invasion and colonisation, all by themselves? Tilman's reply was as follows: it's not that these plants from the neighbouring communities are weaker competitors, on the contrary, many of them are stronger. The reason is different: they have a lower chance of expansion since they were late to arrive to a specific location when the conditions for germination and establishment were more favourable and were overtaken by other species. Thus, according to him, there is a trade-off between the competitive ability and dispersal ability: an organism cannot be both very competitive and capable of a rapid dispersion. It is too soon to say whether similar arguments could explain the enormous plant variety in tropical forests.

Unfortunately, I cannot discuss in this work some alternative opinions, as they are quite numerous. An exception could be made to the so-called Unified Neutral Theory of Biodiversity (UNTB) (Hubbell, 2001). According to the UNTB, the differences between coexisting plant species are insignificant in their functional context, their niches are identical, and their diversity and population size are determined by random speciation and extinction, as well as casual factors, such as wind, water currents, other living organisms responsible for seed dispersal, etc. Hubbell concludes (Ibid. p. 29):

> ... ecological communities are open, continuously changing, nonequilibrium assemblages of species whose presence, absence, and relative abundance are governed by random speciation and dispersal, ecological drift, and extinction.

Ecological drift here is understood as random fluctuations of population sizes which occasionally may lead to a complete elimination from the guild. Some people may see here a similarity with another hypothesis of neutral evolution presented by M. Kimura, J. Crow, J. King and T. Jukes back in 1960s. According to these authors, the high diversity of enzymes in populations revealed by electrophoresis is of no significance and has no impact on adaptivity, and since natural selection cannot influence it, this diversity is nothing more but reflection of stochastic mutagenesis and genetic drift.

The rise of the UNTB has brought back the debate about whether ecological equivalents can be found among coexisting plants. Supporters of this theory sometimes refer to the concept of ecological fitting suggested by Janzen (1985). He developed this concept after his visit to Santa Rosa National Park in Costa Rica. The core of his idea was as follows: species can coexist without eliminating one another even without a long-lasting previous coevolution. For instance, species migrate to more favourable areas because of climate change, however, due to uneven migration rates the communities occurring in new territories are very different although sufficiently stable. This happened also to the tropical rain forests: during the last glaciation, the areas covered by them were much smaller than nowadays, they survived only in a patched and fragmented network of refuges. As the climate has been warming up during the last 16 millennia, these forests have spread, although this process has not come to an end and it might have continued if were not for human interference. Thus, according to Janzen, modern tropical rainforests are temporary communities built from random elements that have not been harmonised by coevolution. Nonetheless, they are pretty stable.

This concept must have been made popular by the story of the Ascension Island. This island emerged from the depths of the Atlantic Ocean about 1 million years ago, at 8° Southern latitude. It is 13.5 km long and 8 km wide, the distance to Brazil is 2200 km, and to Africa 3200 km. Up till the early nineteenth century the island has not had a single tree and resembled a lunar landscape. Only some 20 species of herbs and shrubs were growing there, some of them endemic. In 1815–1817, British sailors planted a garden on the Green Mountain which is 859 m high. The real transformation of the island, however, started only in 1847: over 5000 different plant species had been planted until 1880 (Duffey, 1964), most of them trees and shrubs brought from London and Cape Town botanical gardens. These were plants from almost every tropical region. Already in 1865, the upper part of the mountain was covered by dense tropical forest which was completely unique in terms of its species composition (Catling & Stroud, 2017). Nowadays, one can find some 200–300 introduced species within an area of a dozen square kilometres. If not for the human impact, this ecological community appears to have quite stable species composition.

Some ecologists think that such course of events can be explained only by assuming that even if these plants imported from various locations were compelled to adapt to each other (to share the niches peacefully), they have achieved that in other ways than coevolution (Wilkinson, 2004), as there was not enough time for it. Phenotypic plasticity might have been enough. Does that mean coadaptation without coevolution? This idea can be challenged, however (Gray, 2004): firstly, it seems that nobody has analysed whether during those more than 16 decades the introduced plant populations have experienced any changes of gene frequencies. Such a possibility cannot be rejected, especially bearing in mind that most of the plants brought to the island were propagated by or have undergone their own sexual reproduction, which must have produced some genotype diversity. It seems quite symptomatic in general, that ecologists tend to ignore the role of genetical diversity in the process of adaptation, and not only in this particular case, possibly, because they think that such diversity can only occur by way of mutagenesis and thus requires a lot of time.

By the way, only a small fraction of these 5000 species have become naturalised. The other introduced species failed to germinate or to take root, and have been displaced at the later stages or exterminated by parasites and herbivorous animals. Supposedly, no equivalent species remain among the survivors, or they are very few. This was not the case with all the originally introduced species. However, all these considerations are only speculations because, as it was mentioned, no somewhat more detailed research has been carried out in due time.

It seems the story of the Ascension Island can hardly be interpreted as a proof that the Theory of Neutral Biodiversity offered by Hubbell (2001) is correct. Generally speaking, acceptance of this hypothesis would also reject most of the facts and theoretical constructs (such as the concepts of resource partitioning and complementarity, understanding of competition and the principle of competitive exclusion) that have been accrued by ecologists within the last century and are legitimate objects of their pride. Therefore, many outstanding ecologists (e.g. Fargione et al., 2003; Silvertown, 2004; Levine & HilleRisLambers, 2009), including D. Tilman, use numerous theoretical and empirical arguments to reject most of Hubbell's (2001) assertions.

Despite that, the 'theory' of S. P. Hubbell has attracted enormous attention and generated a high number of supporting and enhancing articles. Even E. O. Wilson seems to have fallen to its charms. What is the explanation?

The first thing that many authors (e.g. Gotelli, 2006), including the opponents of S. P. Hubbell, notice is the similarity of this 'theory' with zero-hypotheses quite common in science, when the researcher poses the question: what would the data look like if the postulated factors (in this case, interaction of species and the resulting resource partitioning) did not exist or had no effect on the examined phenomenon. Such hypotheses are by no means useless, at the very least because they encourage to take another look at the established opinion and review the argumentation. That alone is often enough to make the arguments stricter. On the other hand, they can also be useful (although not necessarily) in the sense that they simplify the problem to the extreme, leaving an idealised core devoid of various aspects or affecting factors. Such simplification allows to develop an uncomplicated mathematical model of the problem and to find simple and internally correct resolutions. In my opinion, Hubbell has chosen the easiest approach by assigning everything to randomness, or rather to unspecified and therefore unmeasurable factors, and thus simplifying the problem. Unfortunately, simple but also effective approaches have been long exhausted in biology. The history of biology has more similar examples, when neat, comprehensible, and convincing mathematical language (just as it is in Hubbell's case) may leave a reader with limited knowledge of scientific methodology with the impression, that everything, or at least quite a lot, can be explained in such an easy and graceful manner. Thus, all the excitement about the 'Neutral Theory' should not come as a surprise. This is not the first instance, by the way. Let us recall, for example, the story of Hardy-Weinberg's law. Its charm lies in the mathematical language easily understood by everyone. A reminder: this law explains what happens to the gene pool of a population when it is not affected by selection, mutations, gene drift, and migration. There might be some who up till now consider this law an efficient instrument that can reveal a lot about microevolutionary processes in nature. Moreover, senior high school students in many countries seem to be still required to know this law, although they have just vague ideas about its applications. It is quite obvious, however, that this law is only an extreme simplification, a description of an ideal state, a zero-hypothesis leading to stricter reasoning, although it is a complete truism to a mathematician and a poor help to a field evolutionary biologist.

One may notice an unpleasant pattern: whenever we, biologists, struggle with a particular problem for a lengthier period, eventually some not very creative but pretty venturesome person would come up with His Majesty Randomness, as the ultimate but reliable tool. Just like a fire, in a flash it sweeps away not only the evidence of a crime but also the crime itself, i.e. the inconvenient problem. As if it has never even existed… 'So, you are asking why? You know, it all boils down to randomness. If we could turn back time, the result might be totally different. It's the butterfly effect…'.

References

Baldeck, C. A., Harms, K. E., Yavitt, J. B., John, R., Turner, B. L., et al. (2013). Soil resources and topography shape local tree community structure in tropical forests. *Proceedings of the Royal Society B, 280*, 1–7.

Brooker, R. W., Maestre, F. T., Callaway, R. M., Lortie, C. J., Cavieres, L. A., et al. (2008). Facilitation in plant communities: The past, the present, and the future. *Journal of Ecology, 96*, 18–34.
Callaway, R. M. (2007). *Positive interactions and interdependence in plant communities*. Springer.
Catling, D. C., & Stroud, S. (2017). The greening of Green Mountain, Ascension Island. In M. Joachim & M. Silver (Eds.), *XXL-XS: New directions in ecological design* (pp. 151–157). ACTAR.
Clements, F. E. (1916). *Plant succession: An analysis of the development of vegetation*. Carnegie Institution of Washington.
Connell, J. H., & Slatyer, R. O. (1977). Mechanisms of succession in natural communities and their role in community stability and organization. *American Naturalist, 111*, 1119–1144.
Davies, S. J., Palmiotto, P. A., Ashton, P. S., Lee, H. S., & Lafrankie, J. V. (1998). Comparative ecology of 11 sympatric species of *Macaranga* in Borneo: Tree distribution in relation to horizontal and vertical resource heterogeneity. *Journal of Ecology, 86*, 662–673.
Duffey, E. (1964). The terrestrial ecology of Ascension Island. *Journal of Applied Ecology, 1*, 219–251.
Ellenberg, H. (1974). *Zeigerwerte der Gefasspflanzen Mitteleuropas*. Goltze.
Ewel, J. J., & Bigelow, S. W. (1996). Plant life-forms and tropical ecosystem functioning. In G. Orians, R. Dirzo, & J. H. Cushman (Eds.), *Biodiversity and ecosystem processes in tropical forests* (Ecological studies) (Vol. 122, pp. 101–126). Springer.
Fargione, J., Brown, C. S., & Tilman, D. (2003). Community assembly and invasion: An experimental test of neutral versus niche processes. *Proceedings of the National Academy of Sciences USA, 100*, 8916–8920.
Flöder, S., Urabe, J., & Kawabata, Z. (2002). The influence of fluctuating light intensities on species composition and diversity of natural phytoplankton communities. *Oecologia, 133*, 395–401.
Gotelli, N. J. (2006). Null versus neutral models: What's the deference? *Ecography, 29*, 793–800.
Gray, A. (2004). The parable of Green Mountain: Massaging the message. *Journal of Biogeography, 31*, 1549–1550.
Hubbell, S. P. (2001). *The unified neutral theory of biodiversity and biogeography*. Princeton University Press.
Hutchinson, G. E. (1961). The paradox of the plankton. *American Naturalist, 95*, 137–145.
Janzen, D. H. (1985). On ecological fitting. *Oikos, 45*, 308–310.
Jordan, C. F. (1985). *Nutrient cycling in tropical Forest ecosystems*. Wiley.
Kelly, C. K., & Bowler, M. G. (2002). Coexistence and relative abundance in forest trees. *Nature, 417*, 437–440.
Kreft, H., & Jetz, W. (2007). Global patterns and determinants of vascular plant diversity. *Proceedings of the National Academy of Sciences USA, 104*, 5925–5930.
Laird, R. A., & Schamp, B. S. (2006). Competitive intransitivity promotes species coexistence. *American Naturalist, 168*, 182–193.
Lekevičius, E., & Loreau, M. (2012). Adaptability and functional stability in forest ecosystems: A hierarchical conceptual framework. *Ekologija (Vilnius), 58*, 391–404.
Levine, J. M., & HilleRisLambers, J. (2009). The importance of niches for the maintenance of species diversity. *Nature, 461*, 254–258.
McIntire, E. J. B., & Fajardo, A. (2014). Facilitation as a ubiquitous driver of biodiversity. *New Phytologist, 201*, 403–416.
Silvertown, J. (2004). Plant coexistence and the niche. *Trends in Ecology and Evolution, 19*, 605–611.
Sommer, U. (1984). The paradox of the plankton: Fluctuations of phosphorus availability maintain diversity of phytoplakton in flow-through cultures. *Limnology and Oceanography, 29*, 633–636.
Sommer, U. (1996). Plankton ecology: The past two decades of progress. *Die Naturwissenschaften, 87*, 293–301.

Sterck, F., Markesteijna, L., Schieving, F., & Poorter, L. (2011). Functional traits determine trade-offs and niches in a tropical forest community. *Proceedings of the National Academy of Sciences USA, 108*, 20627–20632.
Tilman, D. (1977). Resource competition between plankton algae: An experimental and theoretical approach. *Ecology, 58*, 338–348.
Tilman, D. (1986). Resources, competition and the dynamics of plant communities. In M. J. Crawley (Ed.), *Plant ecology* (pp. 51–74). Blackwell Scientific Publications.
Tilman, D. (1994). Competition and biodiversity in spatially structured habitats. *Ecology, 75*, 2–16.
Tilman, D. (1997). Community invasibility, recruitment limitation, and grassland biodiversity. *Ecology, 78*, 81–92.
Tilman, D. (2004). Niche tradeoffs, neutrality, and community structure: A stochastic theory of resource competition, invasion, and community assembly. *Proceedings of the National Academy of Sciences USA, 101*, 10854–10861.
Tilman, D., & Downing, J. (1994). Biodiversity and stability in grasslands. *Nature, 367*, 363–365.
Whittaker, R. H. (1975). *Communities and ecosystems* (2nd ed.). Macmillan Publishing.
Wilkinson, D. M. (2004). The parable of Green Mountain: Ascension Island, ecosystem construction and ecological fitting. *Journal of Biogeography, 31*, 1–4.

Chapter 6
Do Consumers Maintain Diversity of Their Food Sources?

In theory, two extreme opposites are possible. When a herbivore, predator, or parasite is monophagous it can maintain species diversity at the lower trophic level by simply preventing the species it exploits as the food source from excessive reproduction which would otherwise lead to a complete displacement of other species from the guild. As it was mentioned, exploitation in well-established communities is usually moderate (the result of co-evolution), and that means that there should be very few cases of complete extermination of the exploited species in such communities. I shall refer to this method of diversity maintenance as ‘soft’.

Another broadly described mechanism, which I am going to call ‘hard’, is based on the ability of less specialised consumers to shift from the species that has been reduced by their activities to another, more abundant, which otherwise could potentially displace other less competitive species. Unfortunately, in real life, the soft and hard mechanisms are sometimes difficult to distinguish.

In this chapter, I am not going to discuss the cases when consumers get into new unusual locations and encounter food sources that they are not coadapted with. In such cases, both the extinction of the newcomer species and the opposite result, a negative impact on the resident species diversity, are possible. As an example, I could mention some well-known to ecologists facts of the extinction of certain resident organisms (mostly the endemics of oceanic islands) following the introduction of particularly aggressive predators or parasites. Anyway, I shall henceforth focus on the well-established communities only.

Some field experiments have contributed significantly to the conclusion that predators might regulate by the hard way not only the abundance of their prey species but also their diversity, and the classical example thereof is the work of Paine (1966). He isolated small areas within the intertidal zone and removed from them all individuals of the starfish *Pisaster ochraceus*. This marine predator is considered a generalist because it can prey on the filter-feeding mussels *Mytilus californianus*, barnacles *Balanus glandula*, browsing limpets and chitons, as well as feed on dead organisms and single-celled algae. A few months later, the barnacles have spread greatly in the experimental areas, however eventually the mussels have become prevalent as the strongest competitors for food and space. All multicellular algae except one species have become extinct as they lost spatial competition to mussels. Naturally, this took a heavy toll on algae-grazing molluscs. By the end of the experiment, only eight out of fifteen species preyed upon by the starfish were left. Paine concluded that the starfish has become a factor maintaining species diversity in the

E. Lekevičius, *Biodiversity*, https://doi.org/10.1007/978-3-031-11582-0_6

intertidal zone communities because its principal prey were the mussels, the most abundant species. Upon removal of the starfish, the mussels simply prevailed over their weaker competitors. Biodiversity in the control areas with no removals has remained undiminished since the starfish supported the weaker competitors by consuming the mussels and barnacles.

In another one of his numerous experiments, Paine (2002) removed all sea urchins and herbivorous molluscs from the experimental plots on the littoral zone of the eastern US coastline. This move was comparable to the effect produced by sea otters along the western coastline (although sea otters never completely wipe out the sea urchin and mollusc populations). The result was rather unexpected: a diverse brown algae community practically turned into a monoculture of a single species, *Alaria marginata*. The author concluded that without a selective pressure from herbivores focused primarily on *Alaria marginata*, the latter simply outcompetes the weaker species. Strikingly, the productivity of this species (biomass increase within a certain area per unit of time) has grown tenfold in comparison with the productivity in control plots subject to grazing. This indicates that competition must be quite intensive in the guild of algae, and coexistence was secured by the herbivores. It is not very clear, however, whether this is soft or hard regulation of biodiversity.

Porter (1972) demonstrated in a similar way that the crown-of-thorns starfish *Acanthaster planci* which feeds on just one coral species of the *Pocillopora* genus maintains the diversity of the coral guild in the littoral zone of the Panama coast due to its exceptional voracity. The mentioned coral is a very strong competitor capable to eliminate most other coral species from the community when it is not suppressed by the starfish. This case most likely represents the soft regulation.

A study of the bird fauna of nine islands located along the Scandinavian coastline (Kullberg & Ekman, 2000) revealed that four of them support three coexisting tit species: the coal tit, the crested tit, and the willow tit, while the remaining ones are inhabited only by the coal tit. The researchers noted the fact that on the islands with all the three tit species their predator, the pygmy owl, was also present, while it was not detected on the other islands. The authors are inclined to think that on the islands where the pygmy owl is absent the coal tit has simply displaced its weaker foraging competitors. On the other islands, the pygmy owl prevents it from achieving that, because the coal tit comprises a significant share of its prey. This sounds like another example of the hard regulation.

There have been more similar natural or field experiments (see the overviews in Oksanen, 1988; Begon et al., 2006; Krebs, 2009), and most of them support the hypothesis that such 'top-down' control mechanism is widespread in nature and quite efficient. For example, ecologists are well aware (Begon et al., 2006) that foraging niches of the herbivorous mammals are rather broad (with some rare exceptions), therefore they appear to perform the hard control of plant diversity by switching their feeding efforts from one species to another once the biomass of the former is reduced by their activities. Thus, the more competitive and abundant plants are prevented from displacing the rarer ones. Such behaviour is often caused by seasonal changes when certain plants (producers) become suitable or not suitable as food, or when their abundance and productivity fluctuate within very broad boundaries. Obviously, these changes are also related with seasonal migrations of animals.

Where diversity and population densities of large herbivores are high, as in savannah for instance, one may start wondering how is it that they do not undermine plant diversity. The answer may lie in their specialisation. A study on the impact of zebras, Thomson's gazelles, and wildebeests on African plant communities in Serengeti National Park (Gwynne & Bell, 1968) revealed that they exploit the savannah following a rather regular pattern: zebras graze only the tips of the plants, thus opening the access to leaves and leaf sheaths for wildebeests, which, in turn, thin out taller competitive grasses to open space for smaller herbs consumed by the herds of Thomson's gazelles that follow closely. Such selective grazing, when each group of plants has their own consumer and none can become permanently prevalent, must be a diversity-stabilising factor. By the way, a similar selectivity was discovered by D. F. Vesey-Fitzgerald already in 1960 in his study of foraging patterns and hierarchies of hippopotamuses, buffaloes, elephants, zebras, and some smaller herbivores coexisting in the same territory near Lake Rukwa in East Africa.

Already Darwin wrote (1872) that regular mowing can support higher plant diversity in meadows compared to the non-mowed ones. His explanation was very logical: mowing does not allow the more competitive species to overcome the weaker ones. Herbivores seem to have a similar effect on plant diversity. The idea is supported by an experiment performed by Tansley and Adamson (1925, from Begon et al., 2006) almost a century ago. They removed rabbits from areas of species-rich chalk grassland. A few years later, these areas were dominated by a handful of herb species. It turned out that the plants that were favoured by rabbits were also strong competitors, so when the rabbits disappeared, these herbs displaced most other species. However, there are numerous demonstrations (e.g. Zeevalking & Fresco, 1977) that excessive grazing might have a negative effect, except that this time another group, the non-palatable plants, becomes predominant. Of course, such cases are usually associated with human activities.

Mwendera and collaborators (1997) report that the experimental plots of natural pastures where the number of cattle was regulated to ensure moderate grazing, just like in the wild, had much higher numbers of plant species at the end of their experiment, compared to the plots with no grazing at all or a very intensive grazing. The absence of cattle would result not only in the reduction of species numbers but also in domination of certain highly competitive species. Overgrazing, on the contrary, would lead to the prevalence of a few species not suitable for cattle, while most others would be eliminated. However, this effect of diversity maintenance seems to apply only to sufficiently productive plant communities (see Olff & Ritchie, 1998; Bakker et al., 2006; as well as references therein).

One of the most impressive illustrations that predator activities influence not only the diversity of the prey species but also that of plants and other elements of the community is the report on the effects of the reintroduction of wolves into the Yellowstone National Park (north-eastern part of the USA). Back in the nineteenth century, the Park had at least a few-hundred strong wolf population. However, when the US Congress declared the war on them in 1914, they were completely wiped out from the Park by 1926. A few years after the extermination of the wolves, biologists working at the Yellowstone noticed some unexpected trends. For instance, in addition to the significant increase of the red deer population (the main prey of wolves), which could be expected, they noticed substantial changes in the park vegetation cover, especially along the riverbeds. Red deer have mostly damaged their main forage species: willows, young aspens, and poplars. Within a few decades, willows were gone completely, while aspens and poplars were represented only by solitary mature trees. The river valleys have lost their vegetation and demonstrated clear signs of soil erosion. Mature or nearly mature plant communities have been replaced by the plants typical to early succession phases. The populations of trout and some other typical fish have declined in the local streams. Beavers that were once abundant in the Park started leaving it since they had lost their principal sources of food. On the other hand, the coyote population was booming, and they significantly reduced the populations of their main prey, the pronghorn and small rodents. The Park administration was forced to take some drastic measures: at first, red deer were captured and transported to other regions, later, some shooting commenced. The situation has improved somewhat, yet not enough. Finally, the decision was made to reintroduce wolves into the Park, and this was done in 1995. Ten years later, their numbers have approached 200 (a dozen of packs).

Figure 6.1 illustrates the effects of this reintroduction on certain components of the ecosystem. There is no doubt that the plant communities typical of that area are bouncing back after the reintroduction of wolves. The only negative impact that wolves produced was on the red deer and coyote populations: the latter have not only been displaced indirectly, as competitors for the food and territory, but some were simply slain by the wolves. The survivors withdrew to different locations where wolves were less abundant. Bearing in mind the ability of wolves to shift from the prey whose numbers they have diminished towards the more abundant species, we observe another example of the hard regulation.

The Yellowstone lessons should be useful for ecologists for yet another reason: they are a very convincing proof of the importance of indirect relations within ecosystems. Ecological communities are rather deeply integrated systems, therefore any direct impact on one of the components leads to more than just some effect on another one, it reverberates almost as intensely across the entire system and can modify it significantly.

This is further supported by the experience of sea otter decline (Duggins, 1980). The species was once abundant everywhere from Alaska in the north down to California in the south. Its habitat is the brown algae forests, where it preys on different marine organisms including sea urchins, bivalves and gastropods, crustaceans, and fish. These ecosystems provided not only a shelter but also breading and foraging grounds to many animal species. By the end of the nineteenth century, however, sea otters have been almost completely exterminated along this coastline as a valuable fur animal. Only a handful of them managed to survive in some remote areas. The consequences were catastrophic. Without predators to control them, sea urchins have propagated to the levels that almost completely destroyed their own food source, the brown algae and other seaweeds. What followed could be

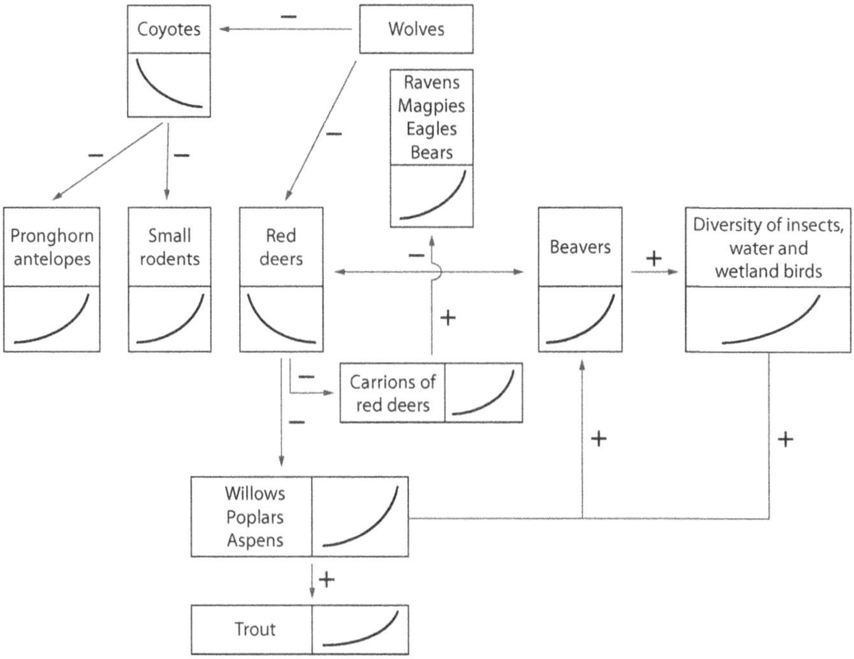

Fig. 6.1 Direct and indirect consequences of the reintroduction of wolves into the Yellowstone National Park. A minus sign next to an arrow indicates negative correlation between the interacting components of the system, while a plus sign – positive one. The upward curves represent an increase of population or species diversity, whereas the declining ones – a decrease. The model was created on the basis of the data taken from Smith et al. (2003), Robbins (2004) and Ripple et al. (2014)

compared to the fate of forest biodiversity after clear-cutting: hundreds of species went extinct, and the marine forests once swarming with life turned into an empty wasteland. Finally, people came back to their senses, and sea otters were reintroduced into the areas where they had been exterminated. The sea urchin numbers went down, and the kelp forests have sprung back to life, so the community is gradually returning to its original state. Once again, we are likely to witness a "hard" regulation of biodiversity performed by sea otters.

Here I would like to remind my readers the article of Hairston and his two co-authors (1960), that has been somewhat forgotten by now but had provoked a heated discussion at the time of its appearance. The authors raised the question why this world is so green. Their suggested answer was as follows: it is green thanks to carnivores and parasites which exploit herbivores; since the former are relatively efficient consumers, the herbivore populations never manage to reach the numbers that could cause essential damage to plants. Should carnivores be eliminated, herbivores would explode to the levels that they would disrupt their own feeding base and perhaps even completely wipe out many plant species. The hypothesis was labelled HSS, according to the initials of the three authors. At first, it was subject to heavy and, in my opinion, quite justified criticism, first of all because it was missing empirical support and represented a coarse oversimplification of the real situation. Nonetheless, it was not rejected completely and, some time later, it was supported by new facts and slightly updated (Oksanen, 1988). Nowadays, most ecologists agree that it is in principle correct. This hypothesis also had much broader implications. According to Oksanen (1988), it might have been the first attempt to demonstrate that 'order can arise from plain Darwinian struggles without ecosystem-level selection'. An even simpler rephrasing would be as follows: according to the HSS, competition and exploitation might be prevailing in nature, but since they act in the opposite directions, they produce surprisingly balanced food webs where elimination of species becomes highly unlikely. In other words, it turned out quite unexpectedly that negative interspecific relations play a positive regulatory role by supporting species diversity, rather than reducing it. By the way, this idea is one of the cornerstones of the ecosystem concept (e.g. Odum, 1971; Patten & Odum, 1981). Sometimes this regulatory effect is referred to as 'exploiter-mediated coexistence', and it is also acknowledged by some experts of population ecology (e.g. Duffy et al., 2007; Berryman & Kindlmann, 2008; Krebs, 2009).

Indeed, there are substantial grounds to assert that α diversity of tropical plants may be determined not only by the variation of the abiotic environment (see Chap. 5) but also by the selective pressure from parasites and herbivores. This is exactly the prediction of the hypothesis presented by Janzen (1970) and Connell (1971). Although it is formulated in rather sophisticated and somewhat confusing manner, I shall attempt to present its key positions in a simple way according to my understanding. Each tropical tree spreads huge amount of seeds, and the number of seeds declines exponentially over the distance from the parental tree. However, the probability of a seed germinating and producing another mature tree is, on the contrary, increasing with the distance from the parental tree. This is due to the fact that each tropical tree species is exploited by numerous specialised herbivores and parasites that can easily migrate from the parental tree onto its progeny growing in proximity. But the spreading abilities of herbivores and parasites are limited, and the further the seed has fallen or been carried by various dispersers, the lower the probability that it or its sprout will be destroyed by the specialised consumers.

This can be presented in an even simpler way. Each parental plant surrounds itself with a rather broad "veto" zone for its own progeny that can reach several hundred meters or more in diameter, where any plant can settle, except for its own species because of the pressure from specialised exploiters. In short, due to consumers, the intra-specific competition becomes a factor that shapes the spatial distribution of progeny not randomly and not according to the seed dispersal trends but in a relatively orderly pattern. This resembles the distribution of pine trees in a pine forest of the temperate zone when the soil patchiness is not high; the only difference is that the distances between mature pines are usually a few hundred times less than those between conspecific plants in a tropical forest. Bearing in mind also the already mentioned (Chap. 5) possible differences according to trophic and abiotic niches, the problem of coexistence of tropical trees does not appear that insolvable. On the other hand, it may be just wishful thinking, according to some opponents.

Some authors attempted to test the Janzen–Connell hypothesis by performing an experiment in the tropical forest environment. By using fungicides and insecticides, they wiped out parasitic fungi (Bell et al., 2006) and herbivorous insects (Bagchi et al., 2014) that attack seeds and sprouts. The survival rate of seeds and sprouts improved, but the species diversity in the cohort of sprouts has declined significantly in comparison with the control plots, just as predicted by the hypothesis (see also Wills et al., 2006).

If asked why, contrary to the tropics, there are almost no limitations to the growth of pine population density in a pine forest, the supporters of the Janzen–Connell hypothesis answer on the following note: that is because in the temperate climate herbivores and parasites are not as efficient regulators of density and diversity. On the other hand, some data suggest that the Janzen–Connell effect can be observed also in temperate forests, except its manifestation is far less apparent there (Packer & Clay, 2000).

Ecologists categorise parasites into specialised and non-specialised ones, or generalists. Contrary to predators, which only consume their prey for food, parasites exploit their hosts also as their habitats to which they must specifically adapt, therefore their degree of specialisation should be much higher (see an overview by Begon et al., 2006).

The specialised parasites seem to be an obvious case: the increase of the host population leads to a higher chance of successful transmission. One would logically expect that, as a very competitive species increases in numbers and densities, the probability of successful transmission of its parasites within the population should also increase. This would result in a negative feedback that slows down further growth of the population. Such development has indeed been discovered in experiments with a grassland plant community (Mitchell et al., 2002). In the experimental plots with artificially reduced plant diversity, the damage caused by specialised pathogens (rust) was much greater than in the plots with higher plant diversity and lower population densities.

Therefore, the relationship between a parasite and its host should be optimised in the course of evolution to attain a compromise between the aggressiveness (virulence) of the parasite, on the one hand, and its transmission efficiency, on the other. In other words, parasites tend to become moderate (Pimentel & Al-Hafidh, 1965; Pimentel & Stone, 1968). It appears quite certain, therefore, that parasites, like other consumers, are capable of regulating not only population densities but, possibly, also diversity of various host species, probably by means of the soft regulation. Even more so, because the biotic potential of parasites, especially of the pathogenic ones, is usually much higher than that of other consumers, thus they are likely to be more efficient regulators.

There is some limited evidence of the possible effect on the diversity of producers caused by parasites that exploit herbivores. For instance, many streams of Michigan, USA, are inhabited by the herbivorous caddisfly *Glossosoma nigrior*. Its larvae feed almost exclusively on attached algae and graze them down to the levels that leave very little to the other stream herbivores. However, occasionally this dominant grazer would be hit by the outbreaks of a highly specific microsporidian parasite *Cougourdella*. Their population shrinks by some two orders of magnitude resulting in the recovery of algae, which has a positive effect on the competitors of *G. nigrior*: they increase both in numbers and species diversity (Kohler & Wiley, 1997).

A similar effect was observed on one of the Caribbean islands in a system consisting of the malarial parasite *Plasmodium azurophilum* and two species of *Anolis* lizards (Schall, 1992). The parasite spreads more easily among individuals of the competitive dominant species and thus prevents it from eliminating the other species. In locations free from the parasite, the weaker competitor is absent.

As for the hard regulation of diversity by parasites, there are significant doubts. Admittedly, as Woolhouse and co-authors (2001; see also Combes, 1996; Begon et al., 2006) noted in their overview, not all parasites are highly specialised, and some of them exploit a few dozens or even over a hundred of host species. For instance, rabies, scabies, or trichinosis can affect hosts representing different orders or even classes. Moreover, the diseases are caused by the same agent species: the *Rabies virus*, *Sarcoptes scabiei*, and *Trichinella spiralis*. Transmission from one host species to another seems highly possible in those cases. After all, this is supported by evidence, which becomes especially strong and broadly known when those agents are transmitted from wild or domestic animals to humans. It appears, however, that such transfer of the so-called generalists is not always possible or at least not fast enough to be suitable for the hard regulation. The reason could be that these parasites often consist of multiple localised populations that could be considered separate varieties or ecological races adapted to one rather than several hosts. So, a species might be a generalist, but its local races are not. The theory does not reject the possibility of transfer under these conditions, but it requires microevolution, and that most likely means that it cannot be fast and efficient in terms of ecological time.

There are some facts, however, that might force us to reconsider this opinion: in the period of 1940–2004 alone, 335 new human parasites have been registered, most of them bacteria (Woolhouse et al., 2005; Jones et al., 2008). Over a half of these cases are zoonoses, i.e. human infections contracted from wild and domestic vertebrates. The total number of human diseases caused by parasites seems to be growing exponentially. No other species appears to have so many parasites. On the other hand, no other species is so aggressively competitive and dominant. This indicates that microevolution of parasites can be much faster than previously considered, especially if the host population has "exploded" to an unprecedented level and density, which facilitates transfers and transmissions. Does that look like a powerful negative feedback created by an overly competitive species and a case of the hard maintenance of biodiversity? It seems to match the definition. However, parasitologists are reluctant to jump to conclusions.

References

Bagchi, R., Gallery, R. E., Gripenberg, S., Gurr, S. J., Narayan, L., et al. (2014). Pathogens and insect herbivores drive rainforest plant diversity and composition. *Nature, 506*, 85–88.

Bakker, E. S., Ritchie, M. E., Olff, H., Milchunas, D. G., Knops, J. M. H., et al. (2006). Herbivore impact on grassland plant diversity depends on habitat productivity and herbivore size. *Ecology Letters, 9*, 780–788.

Begon, M., Townsend, C. R., & Harper, J. L. (2006). *Ecology: From individuals to ecosystems* (4th ed.). Blackwell Publ.

Bell, T., Freckleton, R. P., & Lewis, O. T. (2006). Plant pathogens drive density-dependent seedling mortality in a tropical tree. *Ecology Letters, 9*, 569–574.

Berryman, A. A., & Kindlmann, P. (2008). *Population systems: A general Inroduction* (2nd ed.). Springer.

Combes, C. (1996). Parasites, biodiversity and ecosystem stability. *Biodiversity and Conservation, 5*, 953–962.

Connell, J. H. (1971). On the role of natural enemies in preventing competitive exclusion in some marine animals and in rain forest trees. In P. J. Den Boer & G. R. Gradwell (Eds.), *Dynamics of populations* (pp. 298–312). Wageningen.

Darwin, C. (1872). *The Origin of Species by Means of Natural Selection* (6th ed.). John Murray (Printed in 1998 by Cox & Wyman). Reading.

Duffy, J. E., Cardinale, B. J., France, K. E., McIntyre, P. B., Thébault, E., & Loreau, M. (2007). The functional role of biodiversity in ecosystems: Incorporating trophic complexity. *Ecology Letters, 10*, 522–538.

Duggins, D. O. (1980). Kelp beds and sea otters: An experimental approach. *Ecology, 61*, 447–453.

Gwynne, M. D., & Bell, R. H. V. (1968). Selection of vegetation components by grazing ungulates in the Serengeti National Park. *Nature, 220*, 390–393.

Hairston, N. G., Smith, F. E., & Slobotkin, L. B. (1960). Community structure, population control, and competition. *American Naturalist, 94*, 421–425.

Janzen, D. H. (1970). Herbivores and the number of tree species in tropical forests. *American Naturalist, 104*, 501–508.

Jones, K. E., Patel, N. G., Levy, M. A., Storeygard, A., Balk, D., et al. (2008). Global trends in emerging infectious diseases. *Nature, 451*, 990–993.

Kohler, S. L., & Wiley, M. J. (1997). Pathogen outbreaks reveal large-scale effects of competition in stream communities. *Ecology, 78*, 2164–2176.

Krebs, C. J. (2009). *Ecology: The experimental analysis of distribution and abundance*. Benjamin Cummings.

Kullberg, C., & Ekman, J. (2000). Does predation maintain tit community diversity? *Oikos, 89*, 41–45.

Mitchell, C. E., Tilman, D., & Groth, J. V. (2002). Effects of grassland plant species diversity, abundance, and composition on foliar fungal disease. *Ecology, 83*, 1713–1726.

Mwendera, E. J., Saleem, M. A. M., & Woldu, Z. (1997). Vegetation response to cattle grazing in the Ethiopian highlands. *Agriculture, Ecosystems and Environment, 64*, 43–51.

Odum, E. P. (1971). *Fundamentals of ecology* (3rd ed.). W. B. Saunders.

Oksanen, L. (1988). Ecosystem organization: Mutualism and cybernetics or plain Darwinian struggle for existence? *American Naturalist, 131*, 424–444.

Olff, H., & Ritchie, M. E. (1998). Effects of herbivores on grassland plant diversity. *Trends in Ecology and Evolution, 13*, 261–265.

Packer, A., & Clay, K. (2000). Soil pathogens and spatial patterns of seedling mortality in a temperate tree. *Nature, 404*, 278–281.

Paine, R. T. (1966). Food web complexity and species diversity. *American Naturalist, 100*, 65–75.

Paine, R. T. (2002). Trophic control of production in a rocky intertidal community. *Science, 296*, 736–739.

Patten, B. C., & Odum, E. P. (1981). The cybernetic nature of ecosystems. *The American Naturalist, 118*, 886–895.
Pimentel, D., & Al-Hafidh, R. (1965). Ecological control of a parasite population by genetic evolution in the parasite-host system. *Annals of the Entomological Society of America, 58*, 1–6.
Pimentel, D., & Stone, F. A. (1968). Evolution and population ecology of parasite-host systems. *Canadian Entomologist, 100*, 655–662.
Porter, J. W. (1972). Predation by *Acanthaster* and its effect on coral species diversity. *American Naturalist, 106*, 487–492.
Ripple, W. J., Estes, J. A., Beschta, R. L., Wilmers, C. C., Ritchie, E. G., et al. (2014). Status and ecological effects of the world's largest carnivores. *Science, 343*, 1241484. https://doi.org/10.1126/science.1241484
Robbins, J. (2004). Lessons from the wolf. *Scientific American, 290*(6), 44–49.
Schall, J. J. (1992). Parasite-mediated competition in Anolis lizards. *Oecologia, 92*, 58–64.
Smith, D. W., Peterson, R. O., & Houston, D. B. (2003). Yellowstone after wolves. *Bioscience, 54*, 330–340.
Wills, C., Harms, K. E., Condit, R., King, D., Thompson, J., et al. (2006). Nonrandom processes maintain diversity in tropical forests. *Science, 311*, 527–531.
Woolhouse, M. E., Taylor, L. H., & Haydon, D. T. (2001). Population biology of multihost pathogens. *Science, 292*, 1109–1112.
Woolhouse, M. E. J., Haydon, D. T., & Antia, R. (2005). Emerging pathogens: The epidemiology and evolution of species jumps. *Trends in Ecology and Evolution, 20*, 238–244.
Zeevalking, J. I., & Fresco, L. F. M. (1977). Rabbit grazing and species diversity in a dune area. *Vegetatio, 35*, 193–196.

Chapter 7
Species Diversity Ensures Higher Total Biomass and Helps to Stabilise It

Surprisingly, Darwin was interested in more than just the origin of species, he also tried to answer the question whether higher plant species diversity guarantees also higher total phytomass. He even bothered to carry out a few field experiments. Summing them up, later he wrote (1872, pp. 87–90):

> … the more diversified the descendants from any one species become in structure, constitution, and habits, by so much will they be better enabled to seize on many and widely diversified places in the polity of nature, and so be enabled to increase in numbers ‹…› It has been experimentally proved, that if a plot of ground be sown with one species of grass, and a similar plot be sown with several distinct genera of grasses, a greater number of plants and a greater weight of dry herbage can be raised in the latter than in the former case. The same has been found to hold good when one variety and several mixed varieties of wheat have been sown on equal spaces of ground ‹…› The advantage of diversification of structure in the inhabitants of the same region is, in fact, the same as that of the physiological division of labour in the organs of the same individual body ‹…› No physiologist doubts that a stomach adapted to digest vegetable matter alone, or flesh alone, draws most nutriment from these substances. So, in the general economy of any land, the more widely and perfectly the animals and plants are diversified for different habits of life, so will a greater number of individuals be capable of there supporting themselves.

He then went on (Ibid. p. 88): '…nature follows what may be called a simultaneous rotation'. Amazing insights! Surprisingly, it seems that these Darwin's remarks have never been discussed by modern evolutionary biologists. Possibly, because therein he presented a functional explanation of species diversity, rather than a causal one, which evolutionary biologists are more accustomed to. So, to Darwin, the diversity of genotypes and species was nothing else but a form of specialisation that could contribute to more efficient use of available resources and higher total biomass production. Through divergence, species not only avoid competition but also complement each other in their daily existence.

What about modern theoretical and experimental ecology, what does it say about the impact of species diversity on the total biomass? Among early theoretical works, I should mention Schoener (1971), who upheld the idea that the guilds consisting of organisms with narrow feeding niches should be able to exploit resources more

E. Lekevičius, *Biodiversity*, https://doi.org/10.1007/978-3-031-11582-0_7

completely and efficiently than those comprising generalist species. Norberg and co-authors (2001) present equally solid arguments. In their opinion, an ever-changing natural environment creates conditions for ecosystems to stock themselves with such diversity of species and phenotypes that would allow for substitutions between species. A community would derive no tangible benefits from species diversity in the short-term perspective because it would include a significant portion of sub-optimal species. However, in the long-term perspective, the more species-rich communities should gain a clear advantage and be able to produce higher total biomass in comparison with a monoculture of even the most productive species.

In any case, the ultimate verdict was up to the field experimenters. In principle, their conclusions matched those presented by Darwin and supported by theoretical modellers: as a rule, multi-specific sets produce higher total biomass than monocultures (Naeem et al., 1994; Tilman et al., 1996, 2001; Hector et al., 1999; Tilman, 2001; Huang et al., 2018). Some results are indeed convincing:

> Compared with the average of the single best species in monoculture, our 16-species plots had 39% greater aboveground biomass and 42% greater total biomass on average for 1999 and 2000. Moreover, 16-species plots in 1999 and 2000 had 2.7 to 2.9 times greater aboveground and total biomass than the average for all species in monoculture. (Tilman et al., 2001)

The underlying mechanisms seem quite obvious to these researchers: the needs and niches of coexisting species within a guild differ, the species complement each other, therefore the available resources are used much more efficiently than in a monoculture. The differences extend beyond merely trophic subniches, therefore, regardless of any environmental variations (within normal boundaries), there will always be some species that will find the existing conditions optimal.

Strange as it may sound, ecologists had much more trouble to tackle the question how species diversity influences the stability of production. A hypothesis about a positive correlation between species diversity and functional stability of ecosystems was presented in the middle of the last century. It was coined by the famous father of general systems theory L. Bertalanffy, together with ecologists R. MacArthur, the Odum brothers (E. and H.), R. Pinkerton, R. Margalef, S. J. McNaughton, and others. These researchers held the opinion that species diversity might be useful to a community in a changing environment, because in such circumstances species can substitute for one another: conditions unfavourable to one species might appear favourable to another one, therefore the energy flows and the total biomass of the ecosystem might remain stable within a broad range of environmental conditions (Fig. 7.1). It is not even important whether we discuss variations in space or in time: both cases result in a countless number of ecological niches, and each of them provides optimum living and breeding conditions for different species. Therefore, at a certain time or in a certain location one set of species prevails, while in another time or location the dominant set will be very different.

The hypothesis has become a practically undeniable truth in the late 1960s. This was especially obvious at an international symposium, held in 1969, which was dedicated to discussing the links between species diversity and ecosystem stability.

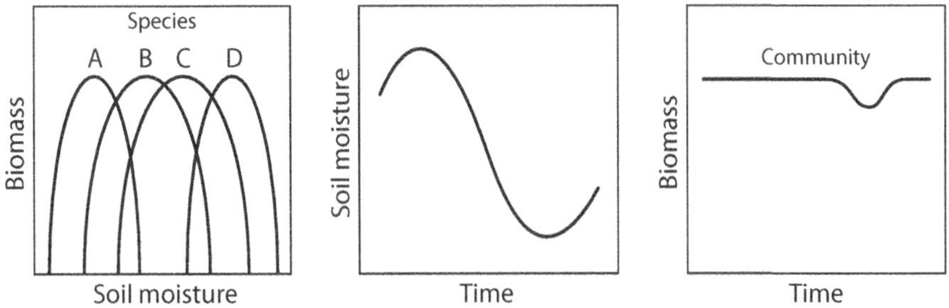

Fig. 7.1 The original hypothesis about the positive correlation between species diversity and community biomass or stability of primary production (based on the example of plant diversity and soil moisture fluctuations). (According to McNaughton (1977), with modifications)

However, as early as in 1974, the attitude changed dramatically. The First International Congress of Ecology was organised in that year, and it was predominated by another group of ecologists supporting the contrary opinion: diversity reduces stability. They have reached this conclusion not by way of observations and experiments as their predecessors, but by way of mathematical modelling. The best-known supporters of this position are F. E. Smith and R. M. May. Later, yet other modellers have demonstrated that even a model can lead to the opposite conclusions, dependent on the initial conditions (e.g. McCann et al., 1998). Not long ago, an elegantly drafted insurance hypothesis was introduced (Yachi & Loreau, 1999; Loreau, 2010). In principle, it is the same original hypothesis, although it includes some new aspects. Among articles trying to build upon the long-lasting experience in this branch of ecology that has witnessed numerous disagreements, I would like to mention the publication of Hooper and 14 other respectful colleagues (2005), which provides the ultimate (?) proof that species diversity adds to the functional stability of a community.

Perhaps some empirical data would be useful to support these conclusions. The best starting point would be something we are already familiar with, for instance, seasonal succession of the plankton in lakes (see Chap. 5), or, more specifically, the relationship between phytoplankton species diversity and functional stability of this guild. At present, it seems to be clear that with seasonal fluctuations of water temperature and nutrient concentrations, different species of phytoplankton gain reproductive advantage, and that, as shown by experiments (e.g. Aruga, 1965; Sommer, 1996), stabilises primary production and the total biomass of phytoplankton. This is understandable: just imagine to what extent, as the weather gets colder, the primary production would drop along with the water temperature in the absence of the cryophilic plankton species. In short, as the water temperature changes, the optimum temperature for plankton photosynthesis fluctuates as well, and optima fitting is achieved through differential reproduction of species.

Upton et al. (1990) analysed total bacteriobenthos of an Antarctic lake. It was cultivated under different conditions: with daily temperature fluctuations from 1 to 16 °C and with a stable temperature (8 °C). Some time later, bacterial diversity was assessed in both experimental sets. It was much higher in the sample with fluctuating regime compared to that of the stable temperature. This sample retained all types of species: the 'cryophilic', the 'thermophilic', and the 'temperate' ones, whereas the second sample included only the 'temperate' bacteria. According to the authors, these results illustrate the importance of the variation of environmental conditions for maintaining species diversity. I can only add my own assumption, that such species diversity should undoubtedly stabilise the activity of total bacteriobenthos from season to season. By the way, one of the species of the *Pseudomonas* genus, which survived in both samples, was subjected to a separate experiment. The bacteria that were pre-adapted to the fluctuating temperature regime were placed together with the bacteria pre-adapted to the stable temperature (8 °C) and then kept under either one (1–16 °C) or another (8 °C) regime for some time. Sooner or later, the preadapted bacteria always displaced the non-pre-adapted ones. This might imply that these two 'ecomorphs' differ also genetically.

Allen and Brock (1968) have come to a similar conclusion. They cultivated a batch of heterotrophic microorganisms composed of samples from various natural sources (topsoil, raw sewage, sewage sludge, etc.) under different temperatures ranging from 2 to 75 °C. It turned out that after a 10-day cultivation period, each experimental temperature will have selected only a small proportion of numerous microorganisms which were pre-adapted to that or another specific temperature. The others will have either disappeared completely, or ceased reproducing, or their reproduction rate will have slowed down dramatically. These experiments might imply that the phenotypic diversity of decomposers in natural environments is so high that their functioning as one of the building blocks of an ecosystem becomes practically independent from environmental conditions (the temperature, in this case).

The ability of bacteriobenthos to adapt its optima to the seasonal temperature fluctuations was demonstrated by Tison et al. (1980). The authors would take bacteriobenthos samples from a lake, then split them into several parts and cultivate each of them under different temperatures in order to establish the preferential, or optimum, temperature for glucose decomposition. It turned out that this temperature changed from 22 °C in summer to 5 °C in winter. It also was discovered that optima fitting was not immediate; it occurred with the delay of about 1.5 months, but the optimum values were almost always higher than the in situ temperature. The authors are convinced that seasonal temperature fluctuation in the lake causes differential reproduction of species, or a succession of the dominant species. The rate of adaptation in this case depends on the fecundity and generation time, and the latter never equals zero. Without a proper species diversity, optima fitting would be impossible: should only relatively thermophilic species exist in the lake, for most of the year the curve of temperature optimum for decomposition of detritus would be well above that of the water temperature near the lake bottom, so decomposition would be very slow.

And on the contrary, in the presence of only cryophilic species, decomposition of detritus would happen only during the cold season, as suggested by the data. A similar result was obtained by Romanenko (1982) as he explored the bacterioplankton of Rybinsk Reservoir. A generalised picture appears as shown in Fig. 7.2.

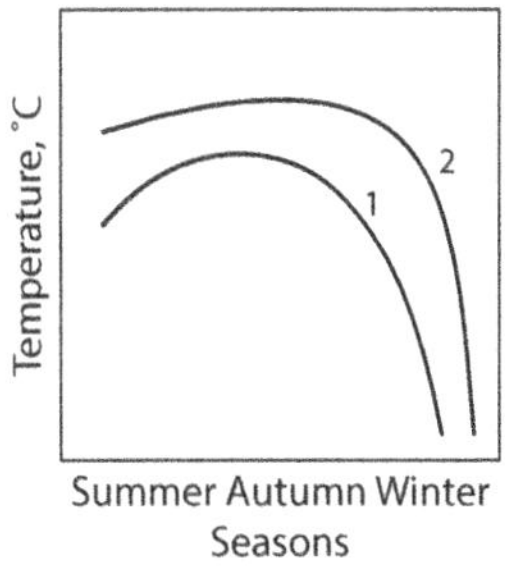

Fig. 7.2 Seasonal changes of the water body temperature (1) and the temperature optimal for bacteriobenthos or bacterioplankton (2). (A generalised picture on the basis of data from Tison et al. (1980) and Romanenko (1982))

Extensive field experiments demonstrated (Tilman et al., 2006) that aboveground plant biomass across all species in a plot fluctuates less from year to year when the number of species is higher. The authors provided the following explanation: the weather conditions of 1 year may be more favourable to one group of species, while those of another year – to another group of species, so different species substitute for each other (and there are less options for substitution in the plots with a lower species richness). The biomass of individual species, however, had a broader fluctuation range in species-rich communities, as compared to the less diverse ones. The explanation could be that the species that lose their optimal conditions under a harsher competition with other species suffer from their competitors more than those in less species-rich communities. One could assume that, if the environmental conditions were stable from year to year, part of the species would be eliminated from the guild. It seems therefore that a usual change of environment is not detrimental to plant diversity, but rather supports it.

In another series of field experiments also involving Tilman (Haddad et al., 2011), an attempt was made to assess how plant diversity affects species diversity at other trophic levels. It turned out that the experimental plots with larger number of plant species also supported more diverse and abundant arthropod communities. The latter included herbivores as well as predators and parasitoids. The authors offer the following explanation of the results: first, plant diversity has a positive impact on primary production and thus provides not only richer food stocks to herbivores but also more habitats suitable for them. The abundance of herbivores attracts predators and parasitoids. Besides, as it turned out, species diversity has contributed to a higher multi-annual stability of diversity which seems to emanate from plants to the higher trophic levels.

What about the role of species diversity in helping a community to adapt to dramatic changes and unusual conditions? Many ecologists have sought the answer to that question. A special attention was drawn by the observations of Tilman and Downing (1994) of the effects of a severe drought, which lasted for 2 years, on prairie communities with various plant diversity levels. In the locations with just a single species growing before the drought, the primary production has diminished 12 times, in those with 3–4 species – about 6 times, whereas in the plots supporting 15–20 species, the decline was only twofold. The relative stability in more species-rich communities was achieved due to broad fluctuations in the numbers of individuals: the populations of more drought-sensitive species had thinned out, but more resistant species had become more abundant. One would expect this work to have finally convinced all the sceptics that the original hypothesis regarding positive correlation between species diversity and functional stability was correct.

Another interesting example in this context is Copeland's (1965) work with an aquarium-based microcosm. When the illumination of pre-balanced marine microcosm (A = R, where A – assimilation, or gross production, of the microcosm and R – microcosm respiration) was reduced more than sixfold, both assimilation and respiration of the community have also significantly diminished within a few days. In the third month of the experiment, however, the assimilation and respiration have attained their original levels (!) and the microcosm has approached a new balance between production and decomposition. The adaptation occurred because turtle grass which had been prevailing before (and which died out because of the insufficient lighting) was replaced by cyanobacteria which are extremely efficient in assimilating dim light. So, in this case, adaptation was possible because the community included not only light-demanding plants but also some shade-tolerant producers. As the conditions changed, it was them that restored the intensity of energy flow. Of course, changes took place not only on the level of producers but also among the consumers. Thus, the laboratory ecosystem has adapted, but it also changed completely in this process and acquired a very different structure.

References

Allen, S. D., & Brock, T. D. (1968). The adaptation of heterotrophic microcosms to different temperatures. *Ecology, 49*, 343–346.

Aruga, Y. (1965). Ecological studies of photosynthesis and mater production of phytoplankton. I. Seasonal changes in photosynthesis of natural phytoplankton. *Botanical Magazine (Tokyo), 78*, 280–288.

Copeland, B. J. (1965). Evidence for regulation of community metabolism in a marine ecosystem. *Ecology, 46*, 563–564.

Darwin, C. (1872). *The origin of species by means of natural selection* (6th ed.). John Murray (Printed in 1998 by Cox & Wyman).

Haddad, N. M., Crutsinger, G. M., Gross, K., Haarstad, J., & Tilman, D. (2011). Plant diversity and the stability of foodwebs. *Ecology Letters, 14*, 42–46.

Hector, A., Schmid, B., Beierkuhnlein, C., Caldeira, M. C., Diemer, M., et al. (1999). Plant diversity and productivity experiments in European grasslands. *Science, 286*, 1123–1127.

Hooper, D. U., Chapin, F. S., III, Ewel, J. J., Hector, A., Inchausti, P., et al. (2005). Effects of biodiversity on ecosystem functioning: A consensus of current knowledge. *Ecological Monographs, 75*, 3–35.

Huang, Y., Chen, Y., Castro-Izaguirre, N., Baruffol, M., Brezzi, M., et al. (2018). Impacts of species richness on productivity in a large-scale subtropical forest experiment. *Science, 362*, 80–83.

Loreau, M. (2010). *From populations to ecosystems: Theoretical foundations for a new ecological synthesis*. Princeton University Press.

McCann, K., Hastings, A., & Huxel, G. R. (1998). Weak trophic interactions and the balance of nature. *Nature, 395*, 794–798.

McNaughton, S. J. (1977). Diversity and stability of ecological communities: A comment on the role of empiricism in ecology. *American Naturalist, 111*, 515–525.

Naeem, S., Thompson, L. J., Lawler, S. P., Lawton, J. H., & Woodfin, R. M. (1994). Declining biodiversity can alter the performance of ecosystems. *Nature, 368*, 734–737.

Norberg, J., Swaney, D. P., Dushoff, J., Lin, J., Casagrandi, R., & Levin, S. A. (2001). Phenotypic diversity and ecosystem functioning in changing environments: A theoretical framework. *Proceedings of the National Academy of Sciences USA, 98*, 11376–11381.

Romanenko, V. I. (Романенко В. И). (1982). Temperature optima of bacterioplankton in Rybinsk reservoir during different seasons of a year. *Microbiology (Moscow), 51*, 866–870. (in Russian).

Schoener, T. W. (1971). The theory of feeding strategies. *Annual Review of Ecology and Systematics, 2*, 369–404.

Sommer, U. (1996). Plankton ecology: The past two decades of progress. *Naturwissenschaften, 87*, 293–301.

Tilman, D. (2001). An evolutionary approach to ecosystem functioning. *Proceedings of the National Academy of Sciences USA, 98*, 10979–10980.

Tilman, D., & Downing, J. (1994). Biodiversity and stability in grasslands. *Nature, 367*, 363–365.

Tilman, D., Wedlin, D., & Knops, J. (1996). Productivity and sustainability influenced by biodiversity in grassland ecosystems. *Nature, 379*, 718–720.

Tilman, D., Reich, P. B., Knops, J., Wedin, D. A., Mielke, T., & Lehman, C. (2001). Diversity and productivity in a long – Term grassland experiment. *Science, 294*, 843–845.

Tilman, D., Reich, P. B., & Knops, J. (2006). Biodiversity and ecosystem stability in a decade-long grassland experiment. *Nature, 441*, 629–631.

Tison, D. L., Pope, D. H., & Boylen, C. W. (1980). Influence of seasonal temperature on the temperature optima of bacteria in sediments of Lake George, New York. *Applied and Environmental Microbiology, 39*, 675–677.

Upton, A. C., Nedwell, D. B., & Wynn-Williams, D. D. (1990). The selection of microbial communities by constant and fluctuating temperatures. *FEMS Microbiology Letters, 74*, 243–252.

Yachi, S., & Loreau, M. (1999). Biodiversity and ecosystem productivity in a fluctuating environment: The insurance hypothesis. *Proceedings of the National Academy of Sciences USA, 96*, 1463–1468.

Chapter 8
Conspecifics Do Not Just 'Struggle' Either

If we were to focus on plants exclusively, we might get the impression of witnessing the 'struggle for existence' in its purest form: the plants of the same population are not bound by any mutual dependence, while their competition is rather intense. But even here we can find some traces of protocooperation in the form of complementation. The old generation of ecological geneticists (e.g. Lewontin, 1974) has always been emphasising the existence of a relatively high diversity of genotypes and phenotypes in natural plant populations, sometimes referred to as genetic polymorphism. This diversity reflects the specialisation of individuals in response to both biotic and abiotic environmental conditions (more on that a little further in this book). It means that the relationship between conspecifics includes an element of complementarity on top of a pure competition. What is also important, this specialisation is transferred across generations and is not just temporary.

The animal world, on the other hand, has many more elements of protocooperation and true cooperation. This has been emphasised by many scholars already since the nineteenth century, some of the most famous names being Espinas (1878), Kropotkin (1902), Wheeler (1923), Allee (1931), Allee et al. (1949), Tinbergen (1953), Wilson (1975), and Berryman and Kindlmann (2008); see also Lekevičius (1986). Quite many of them were convinced that most animals have social organisation. Wilson (1975, p. 7) defines a society as 'a group of individuals belonging to the same species and organized in a cooperative manner'.

Let me try to briefly summarise the works of these researchers. I shall start with the most important phenomenon which is sexual dimorphism. It is an example of widespread true cooperation which often turns into obligatory interdependence. By the way, the same phenomenon can be also observed in the populations of dioecious plants. In this case, the function of sexual reproduction, and 'emergent' quality of a population, is divided between sexual partners that constitute specialised parts of the population. Neither male nor female individuals are able to achieve reproduction without an opposite sex partner, they must enter into an interaction or a contact. It is obvious that both partners are seeking it, although that does not exclude males and females competing for food or other resources.

In most animals, a single individual is not independent not only in terms of reproduction but also in terms of survival. Therefore, conspecifics tend to group together even in cases when the environmental conditions within their distribution range are relatively uniform. Protocooperation is quite common even among most invertebrates. The author of this book has personally witnessed this. While

E. Lekevičius, *Biodiversity*, https://doi.org/10.1007/978-3-031-11582-0_8

experimenting with a *Daphnia* species (Lekevičius & Balčiūnas, 1986), he observed the following phenomenon: while reproducing parthenogenetically, female water fleas produce more offspring (under the same conditions) when they reproduce in groups, together with other females, as opposed to them living in isolation.

A few more examples of protocooperation, among many similar cases: congregations of individuals leading to a more favourable microclimate; schooling of juvenile fish; V-formations of migratory birds; swarms of various organisms, etc.

A family, i.e. parents living together with their immature offspring, should also be considered a form of living in groups which is very common in nature. What relationship prevails in families: competition or cooperation? The only possible answer is 'both'. Parents raising their young usually cooperate and demonstrate altruistic behaviour towards their offspring, whereas the latter are usually completely dependent on the parental care. Relationships between siblings of the same litter are quite different, however, and each sibling acts as a pure egoist, sometimes showing no mercy to the rest of the litter. It is not unusual among large raptor birds that the first hatchling simply ejects the other eggs from the nest or stamps its younger siblings to death. One could hardly find a more illustrative example of the 'struggle for existence'.

Another case of cooperation is colonies created only during the reproduction season. Such colonies are formed by rooks, gulls, cormorants, and many other birds. Although nesting and population densities are extremely high in such colonies, which obviously leads towards increased competition, especially in the periods of food shortage, the shortcomings must be compensated by certain advantages. These are primarily related with the protection of eggs and chicks from predators. Even a fox has difficulties to approach a ground-breeding gull colony undetected – at the first notice, a noisy alarm is raised across the colony and parents invoke mobbing behaviour as a collective defence against the intruder.

Such colonies do not have ranks and hierarchy, all their members being more or less equal in their status. It is 'more or less', however, since older and more experienced individuals usually settle near the core of the colony where a probability of falling victim to a predator is much lower than at the periphery. The latter is left to the younger and weaker. Despite that, colonies are groups of relatively anonymous individuals, where the members either cannot recognise each other 'by face' or do not bother trying, since individual differences are given little importance.

The situation is somewhat different in groups with ranks or hierarchy. Usually, such groups are much more stable than colonies and are retained even beyond the breeding season. These are all sorts of herds, packs, prides, and other groups (e.g. those of crocodiles, monkeys, wolves, lions, stray dogs, etc.). In those cases, individuals, usually the males, fight with each other for a higher rank or status within the group. The winners typically enjoy the privilege of mating and feeding, even if the food is obtained by other members of the group. Severe fighting takes place till the hierarchy is settled, and then it is replaced by the rituals of submissive or subordinate behaviour.

The exceptional feature of these groups is that their members easily recognise each other 'by face' and know which of them have earned obedience and respect and which have not. Besides, these groups demonstrate much higher division of labour than colonies. In some such groups, for instance lion prides, the dominant male is seldom involved in hunting, its main tasks are reproduction and defence against the intruding males. Hunting is done by the other pride members, and even among them there are individuals who prefer to act as scouts and those who draw the game towards the ambush arranged by yet other members, usually strong and agile individuals. Such cooperation makes hunting of a lion pride or a wolf pack much more efficient than that of a solitary individual. Yet other members of the group who are not allowed to reproduce themselves might assume the role of 'nannies', 'aunts', and 'uncles'. Of course, this specialisation is rather provisional and not permanent, so the role of the same individual might change over time.

Even another form of living in groups is true societies or eusocial groups. In addition to the features of the previously mentioned groups, true societies also demonstrate cooperation in construction and maintenance of a common housing, as well as much more advanced communication system for sharing information. Such lifestyle is characteristic to some 2% of insect species (thermites, ants, part of bees and wasps, some beetles, thrips, and aphids), according to Wilson (2008). Some of them also have a caste system, i.e. individuals perform specific roles in their community, determined genetically and thus retained through their entire lifespan. There are other roles that change depending on the age or skills of the individual. For instance, a worker bee usually changes several 'professions' during its lifespan, ranging from housekeeping and nursing, to nest defence, nectar and pollen gathering or scouting. When acting as a scout, a bee is able to use special moves known as 'waggle dance' to transfer the information about newly discovered food source, including distance, direction, and abundance.

Placental mammals also can live in true societies: naked mole-rats, meerkats, prairie dogs, all apes (gorillas, gibbons, orangutans, and chimpanzees), and humans (Allee, 1931; Wilson, 1975). Admittedly, the inclusion of humans into this 'company' is subject to harsh debates, but I do not think they should be extended herein.

Cooperation prevails over competition in such true societies. However, even there some 'struggle' is unavoidable. For instance, the drones in a bee family compete among themselves for the right to fertilise the queen; the first hatched queen kills the other unhatched queens; as a society expands, its members begin to experience food shortage, which increases the tensions and ultimately leads to a part of the society splitting away and colonising a new territory; members of neighbouring conspecific societies often engage in fighting.

Curiously, insect societies seem to have no centralised authority comparable to a dictator, parliament, or government. This distinguishes them significantly from the rank-based groups. Such management type is called diffused, or passive. It is even more amazing, however, how they manage to coordinate activities of numerous individuals engaged together in the construction and repair of relatively complex nests, as well as care and protection of the queen and her progeny. Many aspects of life in these societies remain mystery to us, especially their information exchange mechanisms.

Summing up his understanding of social behaviour in animals, the eminent Tinbergen wrote (Tinbergen, 1953, pp. 114–115):

> When studying the way in which a community (= social groups – E. L.) is organized, one is often struck by the many parallels that can be drawn between it and an individual. Both are composed of constituent parts; the individual is composed of organs, the community of individuals. In both, there is division of labour between the component parts. In both, the parts co-operate for the benefit of the whole, and through it for their own benefit. The constituent partners give and receive. Thus, they lose part of their "sovereignty" as well as part of their capacity to lead a life in isolation. The loss of sovereignty can go so far that parts give their own life for the benefit of the whole. There is constant loss of skin cells in the individual; a lizard's tail is left for the predator for the benefit of the rest of the lizard, so that this rest can live and reproduce. A mother duck defends her chicks even at the cost of her own life. The benefits that the parts derive from the whole is obvious in the individual; an isolated muscle cannot live long. But neither can an isolated worker of the Honey Bee nor an isolated polyp of a Siphonophore colony. Even in cases where individuals can live in isolation, they lose the manifold benefits they receive when living in the flock (...) Loss of the capacity to live outside the community is more striking in the organs of an individual - which has derived its name from it - than in the community; yet the difference is one of degree only. There are individuals which can very well be divided into parts without fatal consequences; tape worms, Planarias and sea anemones are not "undividable".

References

Allee, W. C. (1931). *Animal aggregations: A study in general sociology*. Chicago University Press.

Allee, W. C., Emerson, A. E., Park, O., Park, T., & Schmidt, K. P. (1949). *Principles of animal ecology*. W. B. Saunders Comp.

Berryman, A. A., & Kindlmann, P. (2008). *Population systems: A general inroduction* (2nd ed.). Springer.

Espinas, A. V. (1878). *Des sociétés animales*. Librairie Germer Baillière.

Kropotkin, P. (1902). *Mutual aid: A factor of evolution*. Heinemann.

Lekevičius E. (Лекявичюс Э.) (1986). *Elements of a general adaptation theory*. Mokslas. (in Russsian).

Lekevičius E., Balčiūnas D. (Лекявичюс Э., Бальчюнас Д.) (1986). Polymorphism and complementarity in intraspecific relations: Experiments on Daphnia. *Journal of General Biology (Moscow)*, *47*, 259–267 (in Russian).

Lewontin, R. C. (1974). *The genetic basis of evolutionary change*. Columbia University Press.

Tinbergen, N. (1953). *Social behaviour in animals*. Methuen.

Wheeler, W. M. (1923). *Social life among insects*. Harcourt Brace.

Wilson, E. O. (1975). *Sociobiology. The Abridged Edition*. Belknap/Harvard University Press.

Wilson, E. O. (2008). One giant leap: how insects achieved altruism and colonial life. *Bioscience, 58*, 17–25.

Chapter 9
Genotype Diversity as Even More Delicate Specialisation

Let us start with some important keywords. The following main taxa are distinguished to define heritable differences between conspecifics (in descending order, from major to minor differences):

- Subspecies
- Ecotypes, or ecological races
- Morphs, biotypes, genotypes

Whenever we talk about subspecies, we usually mean distantly separated populations with distinctive differences in appearance or other heritable features. The term ecotype is used by ecologists in reference to the populations with distinctive heritable qualities living under different environmental conditions. For instance, the specimens of Scots pine growing in wetlands appear very different from those growing on the neighbouring higher grounds and these differences are often heritable. Ecotype is also applicable to different populations that are apart and have no contacts with one another. Ecotypes are sometimes referred to as ecological races, forms, or varieties. On the other hand, the terms morph, biotype, or genotype are mainly used to describe minor differences among the individuals of the same population.

Both plant and animal species, especially those with a broad distribution range, usually consist of local populations whose differences are significant enough to consider them different ecotypes or even subspecies. Climatype is yet another term sometimes used in ecology, and it describes the differences between local populations living under different climatic conditions. Climatype is a sort of ecotype. For instance, the Scots pine has 7 subspecies in Eurasia and at least a few dozens of climatypes with distinctive temperature or humidity niches (Rehfeldt et al., 2002). Wide-ranging transplantation experiments have demonstrated that young pine trees transferred into unusual climatic conditions develop less successfully than in their customary environment, and similar results have been obtained with numerous other species. This indicates that climatypes are adapted to particular local conditions and they have been shaped by natural selection. A huge geographical genetic diversity has been discovered among animals, too, however, to the best of my knowledge, nobody has attempted to count the numbers of climatypes. This is understood, as transition from one climatype to another, say, along the latitudinal transect is usually gradual rather than abrupt, therefore the question of the number of climatypes will always remain open and highly subjective. One thing is beyond

E. Lekevičius, *Biodiversity*, https://doi.org/10.1007/978-3-031-11582-0_9

doubt, however: the global number of subspecies and ecotypes (including climatypes), not to mention genotypes, by far exceeds the global number of species.

On the other hand, the status of these subspecific taxa is difficult to define, therefore many researchers use them rather loosely. Some Lithuanian foresters, for example, distinguish a few ecotypes and morphological forms of the Scots pine in their country, while others call many of them varieties, ecological forms, etc. It should be mentioned, for the sake of accuracy, that differences among conspecific individuals are determined not only by genotype but also by phenotype plasticity. Therefore, even very similar genotypes might produce significant phenotype variety under differing conditions, while more different genotypes placed in the same environment produce similar phenotypes. This brings additional complication into taxonomy.

I would like to remind my readers the history of ecological genetics which seems to me unduly forgotten. Even the very title of this scientific subject has by now acquired quite different content which is of little relevance to this monograph (e.g. Conner & Hartl, 2004; Lowe et al., 2004), therefore the achievements of once popular discipline appear even more obscured. I can only regret about such a course of events, without trying to explain the reasons – after all, the issue of preservation of genetic diversity and its adaptive role is too fundamental to be simply forgotten like that. We would hardly find such intellectual swings in, say, physics, chemistry, or mathematics.

Turesson (1923) was the first one to use in his research the method of transplantation which was then improved and has become one of the main tools in plant studies. The seeds of Northern hawkweed (*Hieracium umbellatum*) were collected from distinctively different habitats. They all were planted in a single plot in the researcher's garden. Several generations of the plants then were cultivated in these conditions, and it turned out that in many cases morphological differences between groups ('ecotypes') were preserved in subsequent generations. Turesson assumed that these differences are heritable, and they reflect specific adaptations of the organisms to their local environment.

Subspecies or ecotypes emerge especially easily in cases when local populations are separated by geographical barriers, as these obstruct gene flow (arrival of pollen, spores, or seeds), thus preventing the differences from being levelled out. By the way, it was G. Turesson who introduced the concept of genecology, or genetic ecology, and back then it defined more or less the same field which later was named ecological genetics.

The theoretical basis of the transplantation method is as follows: as it is known, a phenotype, and especially quantitative, or polygenic, traits, depend not only on the genotype but also on the environment, therefore without an experiment it would be difficult to assess whether the observed phenotype differences are heritable or not. If they persist even under the same environmental conditions then they must be heritable, otherwise they are not. To be sure of that, the organisms gathered from different locations are usually kept under the uniform conditions for prolonged periods usually covering at least two generations.

Along with the term ecotype, another broadly used term is polymorphism. The latter was introduced by another founder of ecological genetics, E. B. Ford, who worked with butterflies. In his words (Ford, 1940), polymorphism is:

> the occurrence together in the same locality of two or more discontinuous forms of a species in such proportion that the rarest of them cannot be maintained merely by recurrent mutation.

I am going to use this term in somewhat broader context, as a synonym of the genotype diversity within a population. Ford (1975) has not only introduced the concept of polymorphism; he was also the first to propose an idea that population genetic structure may change in response to environmental changes within the span of one or several generations, thus genetic variation might have a strong effect on demographic processes. In other words, Ford wanted to demonstrate that both time scales (ecological and evolutionary) are not incompatible. This idea became the *credo* of the new discipline.

And yet, for quite a long time, many ecologists and geneticists have been thinking that population genetic structure (gene pool) changes too slowly for these modifications to influence day-to-day activities of communities, demographic changes, and even ecological succession. The turning point in our attitude towards intraspecific diversity was reached in 1966, after the electrophoretic analysis of enzymes had become widely used. It is based on the fact that enzymes with different electric charges travel at different speeds in an electric field. Soon after the first successful trials of this method, thousands of plant and animal populations were analysed. Almost all of them turned out to be polymorphic. Lewontin (1974, p. 118) provides the following summing up of animal research:

> ...we will not be far off if we characterize sexually reproducing species of animals as being polymorphic for a third of their genes, and individuals within the species as being heterozygous for about 10 percent of their loci.

The trend was supported by analogous numbers obtained from plant studies. Thus, although the phenomenon of polymorphism, or genetic diversity, had been discovered way before 1966, only the electrophoretic method finally convinced geneticists that natural plant and animal populations comprise enormous diversity. This radical change was illustrated by R. Levins' book '*Evolution in Changing Environments*' published in 1968, as well as by an international symposium '*Population Biology and Evolution*' organised in the same year by R. C. Lewontin. That year is considered the starting date of ecological genetics. Another outcome of these events was the realisation that genetic processes are not as slow as they were thought to be, and that they participate in the regulation of population size and other dynamic ecological processes. If that is the case, an ecologist must have some knowledge of genetics. The key question of the new discipline has become the diversity of ecotypes and genotypes in populations, its emergence, preservation, and adaptive role.

However, a serious theoretical dilemma arose very soon. You see, back then, the widespread opinion was that populations of all species had to be monomorphic, i.e. they were supposed to consist of just one, the best-adapted genotype which geneticists called the 'wild type'. More genotypes could exist, but only temporarily, as fresh mutants, usually less adapted to their environment than the wild form and therefore doomed to be sooner or later eliminated from the population. Admittedly, in exceptional cases, there could occur a highly successful mutation of such adaptive value to the genotype, that it would surpass even the 'wild type'. In such case, again, the new genotype was expected to sooner or later displace the old one and to become the new 'wild type' itself.

Such opinion was supported by the experts in population genetics who themselves have borrowed it from C. Darwin. The latter perceived natural selection as the force that would each time single out the best version from among all existing ones and brutally eliminate all the others. Lewontin (1974, p. 195) described the situation in this field of biology as follows:

> Classical Darwinian theory held that heritable variations arose from a variety of sources and that natural selection sorted through these variations, rejecting all but the most fit type. Natural selection was seen as *antithetical* to variation. "Many are called, but few are chosen".

So, the facts contradicted Darwinism and Synthetic Theory of Evolution. All the data suggested that:

1. Natural populations are polymorphic.
2. This polymorphism is adaptive in most cases; in other words, it is neither random nor casual.
3. It is not temporary, and the observed alleles seem to be inherited from earlier generations.

This mismatch between the mentioned facts, on the one hand, and the theory, on the other hand, is called the paradox of genetic diversity. How could the situation be resolved? It was not possible to deny the facts since they were too obvious and robust. Adjustment of the theory was the only option. This paradox is linked with the crisis of population genetics in the 1960s and 1970s as well as the advance of ecological genetics in about the same period. The latter, apparently, was more successful in tackling the paradox.

The experts of population genetics proposed two hypotheses (see the overview by Lewontin, 1974). One of them was formulated by Dobzhansky (1955) and, in the same year, by Dempster (1955). It states that there are two forms of natural selection. One of them decreases genetic diversity in populations, and that is a conventional Darwinian selection. The other form increases diversity as it favours heterozygotes that are better adapted than homozygotes. Naturally, a higher degree of heterozygosity should result in high level of polymorphism. This hypothesis refers to a phenomenon of heterosis, well known in genetics. It was named the balance hypothesis (Lewontin, 1974). In the late 1960s, another one, the neutral evolution hypothesis, was presented. Its authors were M. Kimura, J. F. Crow, J. L. King and T. H. Jukes – see Kimura and Ohta (1971). They considered that, in addition to the variation affecting the viability and thus also the adaptive value, there is another, neutral form of variation which is neither bad nor good for the organism. Most variations revealed by electrophoresis are of this type. Therefore, natural selection fails to eliminate them.

Without immersing myself into numerous arguments that had been used while debating these hypotheses, I shall present the main conclusion which was finally reached: both hypotheses match certain facts quite well, however, the remaining and much more numerous facts do not fit into these two schemes. An explanation for the latter was found in the new discipline, ecological genetics, and it boiled down to the following: different genotypes most likely occupy different niches, therefore, as the environmental conditions change in time and space, different genotypes gain advantages or disadvantages, and none of them gets any permanent or absolute advantage. Thus, the 'wild genotype' hypothesis must be rejected since it does not match the facts. This opinion could be called a hypothesis of niche diversity. Its origins are rather old. Already Levene (1953) proposed a hypothesis which stated that polymorphism could be maintained if the environment provides many different sub-niches, rather than just one, and various genotypes have advantages in different sub-niches, which renders them unable to displace each other. Over the next few decades, this idea has been tightened up and elaborated further by other modellers: R. F. Hoekstra, P. W. Hedrick, J. A. Endler, J. H. Gillespie, G. Bell, S. A. Frank, and M. Slatkin. There have been lasting doubts whether environmental changes in time could also have such effect, although some theoreticians, such as J. B. S. Haldane, S. D. Jayakar, R. Levins, S. Ellner, N. G. Hairston, P. Chesson, S. Barot, and J. Gignoux have held a different opinion. They produced models which predicted that not just different genotypes (within a population) but also different species (within a community) can coexist indefinitely, if the fitness of the genotype/species fluctuates in space or time, so that there is always sufficient space or time for that genotype/species to realise its advantages.

Let us try to check whether the facts support these ideas of the modellers. I have already mentioned the by now classical experiments of plant ecotype transplantations performed by Turesson (1923). Nowadays, a more advanced methodology is applied to assess the impact of environment and heredity on phenotype diversity of a population, it is called reciprocal transplantation. Samples of plant species capable of vegetative reproduction are collected from different habitats and cloned to obtain multiple copies of the same genotype. Each clone then is divided into several equal parts and these are planted in different habitats, including the original one wherefrom the clone was taken. During the sampling and transplantation, efforts are made to cause the minimum possible damage to the natural biotope and its vegetation cover. Then all that remains is to observe the development of the experimental plants, their survival rate, biomass, etc. Such experiments in natural environment even allow to quantify the selective pressure working against a particular genotype.

With the help of cloning and transplantation, botanists have discovered huge genetic diversity in the explored plant populations (*Avena, Veronica, Taraxacum, Trifolium, Anthoxanthum, Plantago, Ranunculus, Dryas,* and some other genera), most of which is likely to be the result of microgeographic patchiness of environmental conditions – see the reviews in Nevo (1978, 2001), Bell (1996), Linhart and Grant (1996), Mitton (1997), Ennos (2001), and Kassen (2002).

For example, the studies of a wild oat species – the slender wild oat, *Avena barbata*, which grows in California, enabled Hamrick and Allard (1975) to discover a very delicate specialisation of genotypes. One of the hill tops was occupied by a subpopulation which was monomorphic in terms of the enzyme analysed by electrophoresis. The population at the foot of the hill was also monomorphic but it had the other genotype of the same locus. Finally, the slopes of the hill were characterised by wedge-shaped gradients of the two genotypes: the frequency of the 'summit' genotype increased towards the top of the hill, while that of the 'subjacent' genotype decreased. The authors presumed that such distribution could have been determined by soil characteristics. Subsequent observations in a phytotron, where both genotypes were cultivated in substrates with different moisture levels, supported this hypothesis.

Zoologists have also discovered rather high polymorphism in wild animal populations, wherein genetical structure of certain populations/subpopulations would often differ significantly from the others. A conclusion was made that these differences are determined by microgeographic patchiness of abiotic and/or biotic conditions, and some laboratory experiments even made it possible to identify specific selection factors – see the overviews in Krimbas and Powell (1992), Mitton (1997), Nevo (2001), and Kassen (2002).

For example, Etges (1984), who studied *Drosophila robusta* populations of Great Smoky Mountains, established that the frequencies of chromosomal inversions demonstrate regular changing patterns according to the altitude above the sea. Certain inversions are prevailing at low altitudes, others at medium altitudes, and yet others high in the mountains (some similarity with the example of *A. barbata*). Ananina et al. (2004) discovered inter- and intra-populational diversity of chromosomal inversions within *D. mediopunctata* populations resident in Parque Nacional do Itatiaia (south-eastern Brazil). Their frequencies were also changing in rather orderly fashion along the altitude. According to the authors, some portion of this polymorphism is supported to a large extent by the microgeographic patchiness of environmental conditions.

Schmidt and Rand (1999) studied a northern species of acorn barnacles (*Semibalanus balanoides*), or, to be more precise, their mannose phosphate isomerase (Mpi) genotypes. They found 3 genotypes in the examined population, called SS, SF, and FF. Each genotype has its own microhabitat wherein it is most common. Different genotypes prefer different microhabitats in the intertidal zone: the FF genotype, as demonstrated by transplantation, better survives in the upper intertidal zone affected by higher temperatures, while the SS genotype – in the lower intertidal zone. The authors discovered that this differentiation is preserved across generations, despite the fact that these crustaceans can travel tens and even hundreds of kilometres while in the larva stage. A strong selection pressure must be acting against the establishment of an 'alien" genotype in a particular habitat.

In theory, animals, contrary to plants, can benefit from the tactics of active choice, which is the opposite of passive submission to selection. Obviously, most animals can move and choose the most favourable environment. If a diversity of genotypes does exist in any given population, these genotypes are very likely to disperse in space not randomly but according to the law of maximum adaptedness. There is nothing bad about new genotypes occurring in each generation as the result of gene flow, recombination, and mutation processes since each of them will have at least a theoretical opportunity to find its own optimal environment. Being able to move in space, animals seem to slide along the axes of their ecological niche in both directions in search of the most favourable environmental conditions enabling them to enjoy the maximum adaptedness. Such adaptation based on behavioural reactions and genotype diversity should decrease the genetic load, i.e. the frequencies of poorly adapted genotypes within the population.

It seems that animals are indeed using this opportunity. Some authors (Powell & Taylor, 1979) call this phenomenon an 'intraspecific habitat choice'. It has been observed in marine molluscs *Littorina mariae*, fruit flies *Drosophila persimilis, D. simulans, D. pseudoobscura*, peppered moth *Biston betularia*, and some other species (Powell & Taylor, 1979; Jaenike & Holt, 1991). The case of the peppered moth is well known to most of us from school textbooks, although many readers might have omitted the fact discovered by H. B. D. Kettlewell, that different morphs of this moth demonstrated different behaviour while perching on a tree stem: the pale ones were seeking light spots to be less conspicuous to the insectivorous birds, while the dark ones were bustling around in search of the darkest possible background.

Here I would like to mention also the data obtained by Christensen (1977). He worked with water louse *Asellus aquaticus* (Isopoda), a common species in our waters. The author had chosen a small pond as his study area, with a willow tree growing on one shore and a beech on the opposite. This is an important detail, as the rotten tree leaves comprise a significant part of these isopods' diet. With the help of electrophoretic method, Christensen identified two genotypes in the population that had different types of the enzyme amylase. One genotype was prevailing in the 'willow' side of the pond, while the second – in the 'beech' one. The author assumed that each genotype selects the optimal side of the pond offering the most suitable food. The researcher then conducted an experiment to test his hypothesis. He arranged an aquarium, littering its bottom with willow leaves at one end and beech leaves at the other, then filled it with water and released there some water lice of both genotypes. At first, the genotype distribution was completely random, devoid of any noticeable order. In a fortnight, however, the analysed population has split into two subpopulations rather distinctive in terms of the amylase genotype, just as predicted by the author. Since amylase is involved in food digestion, Christensen was able to see not just a mere correlation but also cause and effect relationship between the locations of food sources and spatial distribution of genotypes.

Genotype diversity seems to be supported not only by the diversity of abiotic factors but also by the biotic environment. Edmunds and Alstad (1978) discovered an amazing relationship between the ponderosa pine and the population of scale insects that are feeding on this pine. Each time when the researchers would capture a sample of scale insects foraging on the needles of one pine tree and place them onto another tree of the same species, the growth and reproduction of the insects would be impaired. This suggested that the insects were well adapted to their particular host tree but less so – to its other conspecifics. A suspicion arose that individual pine trees may differ by their natural chemical compounds which determine their suitability for the insects. Later, it indeed turned out that populations of that pine species are polymorphic according to the composition and content of their monoterpenes. These compounds determine the attractiveness, or rather the repulsiveness, of the plant to its consumers. On the other hand, genetic diversity in pine population stimulates diversity in the scale insect population and its branching into subpopulations which become specialised to exploit the plants they can cope with more easily. In general, as demonstrated by various studies performed by numerous authors over many years, populations of herbivorous insects are highly polymorphic according to the genes that determine their feeding needs. This is the explanation of a well-known fact that, after a relatively short adaptation involving rapid propagation of certain morphs and extinction of others, insects can rather easily switch to a new, unusual food source (Kamshilov, 1961).

Martin and Harding (1981) carried out an extravagant experiment with two *Erodium* (heron's bill) species. These two species grow together (sympatrically) in certain locations and separately (allopatrically) in some others. The authors collected samples of these plants from the wild and cultivated them in a room under controlled conditions. Two sets were used in the experiment: a mixed group consisting of individuals of both species that were growing together in the wild, and a mixed group made from allopatric individuals of both species. The mass of seeds at the end of the vegetation period was chosen as a measure of interspecific competition. It was higher in the first set, implying a weaker competition. The authors concluded that, as species grow together, competition creates a selective pressure which favours the individuals (and corresponding genotypes) with clearly distinctive needs. Such reduction of the overlap of ecological niches must soften competition.

Turkington and Harper (1979) have also addressed this problem. They aimed to check whether different specimens (i.e. genotypes) of *Trifolium repens* (white clover) growing in different biotic environments are better adapted to their particular environment than to a foreign one or is their presence in a particular habitat merely a coincidence. For that purpose, the authors collected white clover samples from 4 communities with different gramineous plants prevailing:

1. Common bent-grass, *Agrostis tenuis* (79.6% of all plants)
2. Crested dog's-tail, *Cynosurus cristatus* (56.5%)
3. Common velvet-grass, *Holcus lanatus* (94.6%)
4. Perennial rye-grass, *Lolium perenne* (96.4%)

The method of reciprocal transplantation was used: the genotypes of white clover were cloned, then planted again in the wild, each genotype placed into 'its own' community and into the three 'foreign' ones. A year later, the transplants were collected to assess their dry weight. It turned out that the plants demonstrated the best growth when planted in their 'own' community.

Vavrek (1998) performed a similar experiment with different genotypes of the common dandelion, *Taraxacum officinale*, sampled from the wild. Field observations suggested that the studied dandelions were usually growing in the company of the following competitors: *Plantago major*, *Poa pratensis*, and *Trifolium pratense*. In certain locations, all the four species were present, while in others there were only some of them. Since all these species follow seasonal vegetation cycles, the interactions between competitors should variate also in time. Vavrek, therefore, selected a few genotypes of dandelion, cloned them, and planted the replicas in a greenhouse together with one or another of the competitor species. Different genotypes demonstrated different reactions towards competitors: while certain genotypes were growing better amidst *P. major* communities, some others were better-off growing next to *P. pratensis* or *T. pratense*. According to the author, his data most likely indicate that different reactions of genotypes to the biotic environment, which fluctuates in time and space, might help maintain genetic diversity within a population.

Prati and Schmid (2000) presented similar results obtained from their experiments with the creeping spearwort (*Ranunculus reptans*). Part of the individuals of the studied population were growing at the water edge with practically no competitors around them (the authors gave them the nickname of the 'lake' genotype) while the other part was spread for up to 10 meters away from the water (the so-called 'land' genotype). The latter were usually found in communities with the creeping bent-grass (*Agrostis stolonifera*). The authors sampled 16 plants from each set and cloned them, then planted the progeny into the plots with different biotic environment set up to imitate the real biotic differences in the wild. The experiment demonstrated that the clones were thriving best in their 'native' habitats: the 'lake' genotype grew better in the non-competitive environment, while the 'land' one – under the conditions of interspecific competition. The authors were convinced that interspecific competition had become an effective selection factor which resulted in the observed split of the population into two subpopulations with different competitiveness levels.

As one can see, numerous studies suggested that the adaptive value of a genotype is usually the highest in its 'native' habitat. This is a likely indication that the studied genotypes are specialised towards the environmental conditions existing within the population's distribution range. There is no doubt that this sort of genotype differentiation in plants is both invoked and maintained by natural selection. Admittedly, a genotype suitable for that or another micro-geographical environment often reaches its 'proper' micro-niche purely by chance, still as a seed, therefore intense natural selection within each generation and a relatively high genetic load are inevitable. Nonetheless, it seems that many, if not most, micro-niches suitable for the population are usually colonised by specialised genotypes. As illustrated by these studies, a gene pool is a kind of storage facility for the information on the diversity of environmental conditions across the distribution range. Curiously, whenever environmental conditions differ significantly between adjacent habitats, the differentiation is often preserved despite rather intense gene flow between the local subpopulations (e.g. Antonovics et al., 1971).

Strictly scientific approach, however, required a direct proof that genetic diversity is determined specifically by the diversity of environmental conditions. Such a proof could only be provided by crucial experiments. So, they were carried out. For instance, Powell (1971) captured sufficient amount of fruit flies *Drosophila willistoni* in the wild and placed them into boxes constructed especially for that purpose. Some boxes contained an environment which was stable in time and uniform across the space. In some other boxes, one of the three factors was subject to changes: the substrate, the food (both in terms of their spatial location), or the temperature (in time). A few more boxes had all the listed factors fluctuating. In 45 weeks, every fly was examined for 22 loci by electrophoresis. The highest genetic diversity was found in the boxes where all the factors were fluctuating, and it was almost as high as in the wild population. In the stable environment, polymorphism has decreased more than twofold. In their later experiments with *D. pseudoobscura*, Powell and Wistrand (1978) came up with the same conclusion.

Another team of researchers pursuing similar goals (Rainey & Travisano, 1998) experimented with the bacterium *Pseudomonas fluorescens*. A heterogeneous medium (no shaking of the microcosm) supported a stable coexistence of at least three morphs (genotypes), however when the medium was made homogeneous (by constantly shaking the microcosm), only one of the morphs would eventually remain. Whenever a heterogeneous environment was restored (no shaking), genotype diversity would increase again, presumably, by way of mutations and selection. The authors had every reason to assume that this 'adaptive radiation' which took just one week to happen was provoked by the occurrence of unoccupied niches (see also Kassen, 2002; Kassen & Rainey, 2004).

I think the data presented in this chapter provide a rather convincing demonstration that genotype diversity is a way to split the niches of a population into subniches, which is quite common in nature. It is also a way to add complementarity to the competitive intraspecific relations with the help of disruptive selection (Levene, 1953). Thus, it would be quite normal to think that genotype diversity might contribute to a more efficient use of resources, which should result in higher reproduction rates, productivity, or total biomass in comparison with the monomorphic populations. Darwin also wrote that mixtures of varieties were usually more productive than pure cultures (see Chap. 7). His opinion was confirmed by subsequent laboratory and field experiments (e.g. Mather, 1961; Tošić & Ayala, 1981; Lekevičius & Balčiūnas, 1986; Crutsinger et al., 2006; Hughes et al., 2008). Together with other material presented above, this fact should be interpreted as a serious testimony with the ambition of functional explanation of genotype diversity.

References

Ananina, G., Peixoto, A. A., Bitner-Mathé, B. C., Souza, W. N., Silva L. B. da, et al. (2004). Chromosomal inversion polymorphism in *Drosophila mediopunctata*: Seasonal, altitudinal, and latitudinal variation. *Genetics and Molecular Biology, 27*, 61–69.

Antonovics, J., Bradshaw, A. D., & Turner, R. G. (1971). Heavy metal tolerance in plants. *Advances in Ecological Research, 7*, 1–85.

Bell, G. (1996). *Selection: The mechanism of evolution*. Chapman and Hall.

Christensen, B. (1977). Habitat preference among amylase genotypes in *Asellus aquaticus* (Isopoda, Crustacea). *Hereditas, 87*, 21–26.

Conner, J. K., & Hartl, D. L. (2004). *A primer of ecological genetics*. Sinauer Associates.

Crutsinger, G. M., Collins, M. D., Fordyce, J. A., Gompert, Z., Nice, C. C., & Sanders, N. J. (2006). Plant genotypic diversity predicts community structure and governs an ecosystem process. *Science, 313*, 966–968.

Dempster, E. R. (1955). Maintenance of genetic heterogeneity. *Cold Spring Harbor Symposium on Quantitative Biology, 20*, 25–32.

Dobzhansky, T. (1955). A review of some fundamental concepts and problems of population genetics. *Cold Spring Harbor Symposium on Quantitative Biology, 20*, 1–15.

Edmunds, G. F., & Alstad, D. N. (1978). Coevolution in herbivores and conifers. *Science, 199*, 941–945.

Ennos, R. A. (2001). Inferences about spatial patterns in plant populations from the analysis of molecular markers. In J. Sivertown & J. Antonovics (Eds.), *Integrating ecology and evolution in a spatial context* (pp. 45–71). Blackwell Science Ltd..

Ford, E. B. (1940). Polymorphism and taxonomy. In J. Huxley (Ed.), *The new systematics* (pp. 493–513). Clarendon Press.

Ford, E. B. (1975). *Ecological Genetics* (4th ed.). Chapman and Hall.

Hamrick, J. L., & Allard, R. W. (1975). Correlations between quantitative characters and enzyme genotypes in *Avena barbata*. *Evolution, 29*, 438–442.

Hughes, A. R., Inouye, B. D., Johnson, M. T. J., Underwood, N., & Vellend, M. (2008). Ecological consequences of genetic diversity. *Ecology Letters, 11*, 609–623.

Jaenike, J., & Holt, R. D. (1991). Genetic variation for habitat preference: Evidence and explanations. *American Naturalist, 137*, S67–S90.

Kamshilov M. M. (Камшилов М. М.) (1961). *The importance of interrelations of organisms in evolution*. USSR Academy of Sciences Publ. (in Russian).

Kassen, R. (2002). The experimental evolution of specialists, generalists, and the maintenance of diversity. *Journal of Evolutionary Biology, 15*, 173–190.

Kassen, R., & Rainey, P. B. (2004). The ecology and genetics of microbial diversity. *Annual Review of Microbiology, 58*, 207–231.

Kimura, M., & Ohta, T. (1971). Protein polymorphism as a phase of molecular evolution. *Nature, 229*, 467–469.

Krimbas, C. B., & Powell, J. R. (1992). Introduction. In C. B. Krimbas & J. R. Powell (Eds.), *Drosophila inversion polymorphism* (pp. 1–52). CRC Press.

Lekevičius E., Balčiūnas D. (Лекявичюс Э., Бальчюнас Д.) (1986). Polymorphism and complementarity in intraspecific relations: Experiments on Daphnia. *Journal of General Biology (Moscow), 47*, 259–267 (in Russian).

Levene, H. (1953). Genetic equilibrium when more than one ecological niche is available. *American Naturalist., 87*, 331–333.

Lewontin, R. C. (1974). *The genetic basis of evolutionary change*. Columbia University Press.

Linhart, Y. B., & Grant, M. C. (1996). Evolutionary significance of local genetic differentiation in plants. *Annual Review of Ecology and Systematics, 27*, 237–277.

Lowe, A., Harris, S., & Ashton, P. (2004). *Ecological genetics: Design, analysis, and application*. Blackwell Publ.

Martin, M. M., & Harding, J. (1981). Evidence for the evolution of competition between two species of annual plants. *Evolution, 35*, 975–987.

Mather, K. (1961). Competition and cooperation. In *Mechanisms in biological competition. Symposia of the Society for Experimental Biology* (Vol. XV, pp. 261–281). New York.

Mitton, J. B. (1997). *Selection in natural populations*. Oxford University Press.

Nevo, E. (1978). Genetic variation in natural populations: Patterns and theory. *Theoretical Population Biology, 13*, 121–177.

Nevo, E. (2001). Evolution of genome – Phenome diversity under environmental stress. *Proceedings of the National Academy of Sciences USA, 98*, 6233–6240.

Powell, J. R. (1971). Genetic polymorphisms in varied environments. *Science, 174*, 1035–1036.
Powell, J. R., & Taylor, C. E. (1979). Genetic variation in ecologically diverse environments: Environmental diversity and intraspecific habitat choice, rather than heterosis, may be the answer to the question "why so many genotypes?". *American Scientist, 67*, 590–596.
Powell, J. R., & Wistrand, H. (1978). The effect of heterogeneous environments and a competitor on genetic variation in Drosophila. *American Naturalist, 112*, 935–947.
Prati, D., & Schmid, B. (2000). Genetic differentiation of life-history traits within populations of the clonal plant *Ranunculus reptans*. *Oikos, 90*, 442–456.
Rainey, P. B., & Travisano, M. (1998). Adaptive radiation in a heterogenous environment. *Nature, 394*, 69–72.
Rehfeldt, G. E., Tchebakova, N. M., Parfenova, Y. I., Wykoff, W. R., Kuzmina, N. A., & Milyutin, L. I. (2002). Intraspecific responses to climate in *Pinus sylvestris*. *Global Change Biology, 8*, 912–929.
Schmidt, P. S., & Rand, D. M. (1999). Intertidal microhabitat and selection at Mpi: Interlocus contrasts in the northern acorn barnacle, *Semibalanus balanoides*. *Evolution, 53*, 135–146.
Tošić, M., & Ayala, F. J. (1981). Density and frequency-dependent selection at the Mdh-2 locus in *Drosophila pseudoobscura*. *Genetics, 97*, 679–701.
Turesson, G. (1923). The scope and import of genecology. *Hereditas, 4*, 171–176.
Turkington, R., & Harper, J. L. (1979). The growth distribution, and neighbour relationships of *Trifolium repens* in a permanent pasture. IV. Fine-scale biotic differentiation. *Journal of Ecology, 67*, 245–254.
Vavrek, M. C. (1998). Within-population genetic diversity of *Taraxacum officinale* (Asteraceae): Differential phenotype response and effect on interspecific competition. *American Journal of Botany., 85*, 947–954.

Chapter 10
Environmental Fluctuations and the Daily Role of Selection

As mentioned at the beginning of Chap. 9, for quite some time many modellers have been doubting whether fluctuations of certain parameters in time can have a similar impact on genotype diversity as spatial patchiness of the environment. Eventually, however, the belief that such impact is possible has gained support. Of course, empirical data contributed to that. For instance, Dobzhansky (1943, 1947) described polymorphism of chromosome inversions (ST, CH, and AR) in the *Drosophila pseudoobscura* populations of the USA. Different morphs prevail in different seasons, and this variation of frequencies follows an orderly pattern from year to year. The author was convinced that this is the result of natural selection, whereby populations demonstrate a trend to adapt to a particular season. A natural assumption would be that this polymorphism might be supported by seasonal fluctuations of the environmental conditions. However, the author did not go far enough to pinpoint the specific factors that may be responsible for this selection: the temperature, humidity, or something else. Ford (1975) labelled this form of selection 'cyclic selection', but the term 'seasonal selection' seems to be more appropriate.

Later, seasonal fluctuations of chromosome inversion frequencies have been discovered also in other studied *Drosophila* species, albeit not in all (Krimbas & Powell, 1992; Ananina et al., 2004; Rodriguez-Trelles et al., 1996). Then again, numerous attempts to establish whether seasonal environmental fluctuations could maintain also allo- and isozyme polymorphism in *Drosophila* quite often showed negative or ambiguous results (e.g. Dobzhansky & Ayala, 1973; Mueller et al., 1985). Yet other researchers (Bergland et al., 2014) managed to detect seasonal selection in fruit flies with the help of the single-nucleotide polymorphism (SNP) method. They suggest that these reorganisations have an adaptive character.

Especially valuable results were obtained by King (1972), who studied the rotifers *Euchlanis dilatata* living in one of the US lakes. He would sample the rotifers at biweekly intervals, then would separate individuals in the laboratory and allow them to produce parthenogenetic clones at the temperatures of 19 °C and 27 °C. It turned out that the genotypes of spring population reproduced faster at 19 °C, whereas the genotypes prevailing in the lake in summer reproduced much faster at 27 °C than at 19 °C. The results most likely indicate that either 'cryophilic' or 'thermophilic' genotypes gain advantage in different seasons, and this is reflected in their frequencies of occurrence. King also proved that physiological adaptation to the changing temperatures (reorganisation within the boundaries of individual reaction norm) also takes place, along with the differential reproduction of genotypes.

E. Lekevičius, *Biodiversity*, https://doi.org/10.1007/978-3-031-11582-0_10

Carvalho (1987) performed similar experiments on a wild population of *Daphnia magna*. He discovered that electrophoretically distinct clones prevail in the lake in different seasons. Laboratory analysis revealed that ‘thermophilic’ clones dominate in summer, while ‘cryophilic’ in winter, and the difference between their temperature preferenda is predetermined genetically and reaches 5–10 °C.

Comparable observations were carried out by Gurevičiūtė et al. (1989; see also Lekevičius, 1997) who were working with a wild population of *Daphnia pulex*. The methodology used was similar to that of King (1972). The *Daphnia* population demonstrated genotype diversity according to the dependency of parthenogenetic reproduction rates on temperature: some genotypes were relatively ‘thermophilic’ and some others ‘cryophilic’. Most genotypes sampled in summer were still ‘cryophilic’, and the prevalence of the ‘thermophilic’ ones became noticeable only since the beginning of September, although the water temperature was already declining by that time. A more detailed investigation revealed that the maximum frequencies of the ‘cryophilic’ genotypes have a lag of about 1–1.5 months in comparison with the temperature curve. The authors came up with the following explanation: even though certain genotypes lose their optimum conditions, they are unlikely to disappear immediately; as different clones that were scarce until then gain advantage, it takes time for them to propagate to the level when they become prevalent, thus increasing the level of adaptedness of the entire population. The reproduction rate, and thus the rate of adaptation, is restricted by the fecundity and generation time which, in the case of *Daphnia,* is at least 5–8 days. This seasonal adaptation shows a similar success rate as a comparable adaptation of entire guilds by way of seasonal succession (see Fig. 7.2).

Niklasson et al. (2004) presented data on the genetic structure changes of a spear-winged fly *Dipsa bifurcata* found in southern Sweden over the span of 8 years. Several populations were analysed, and recurring seasonal and year-to-year fluctuations of clone frequencies were discovered in all of them. The authors consider that this phenomenon is caused by fluctuations of climatic factors. Seasonal fluctuations of clone frequencies have also been discovered in *Heterocypris incongruens* (Ostracoda; Rossi & Menozzi, 1990) and one of the tick species (Weeks & Hoffmann, 1998).

These works demonstrate that differential reproduction of genotypes, which is nothing else but natural selection, can play a very important role in adaptation to seasonal phenomena. Of course, such seasonal adaptations through natural selection are only possible in organisms whose generation time is no longer than 2–3 weeks (Lekevičius & Loreau, 2012). Otherwise, the reaction lag would be so great that their benefits to the population would be doubtful. Even *Daphnia* species hardly manage to keep up changing the genetic structure of their population at such a rate that population reproduction is almost undisturbed with the arrival of autumn. Thus, more advanced crustaceans, molluscs, fish, and other large animals, as well as plants, are out of question. These groups of organisms can adapt to seasonal phenomena only by way of individual mechanisms. Natural selection would be too slow for that purpose.

However, genotype diversity in the populations of these organisms seems to be supported by year-to-year fluctuations of environmental conditions. Such possibility is suggested, for instance, by the data obtained by Kelly et al. (2003) from the studies of genotype diversity in the common birch, *Betula pendula*. These authors demonstrated that the analysed population consisted of relatively 'thermophilic' and 'cryophilic' genotypes. In different years, natural selection favours different genotypes within the youngest cohorts (the youngest age groups); in warmer springs, the first group enjoys better germination and viability of sprouts, while colder springs are better for the second group. Since sprout survival is the critical stage in the life cycle of plants when they are especially vulnerable to climatic conditions, the surviving saplings become more resistant to endure even a less favourable year. The ultimate result should be such that certain cohorts are dominated by the 'thermophilic' while others by the 'cryophilic' genotypes (Fig. 10.1).

In this case, we do not observe annual changes of genotype frequencies, they occur only within the boundaries of individual cohorts. However, such variation of climatic conditions might be quite sufficient to ensure stable coexistence of 'thermophilic' and 'cryophilic' genotypes. One could expect that without such climatic fluctuations one genotype would eventually displace the other one. In this example,

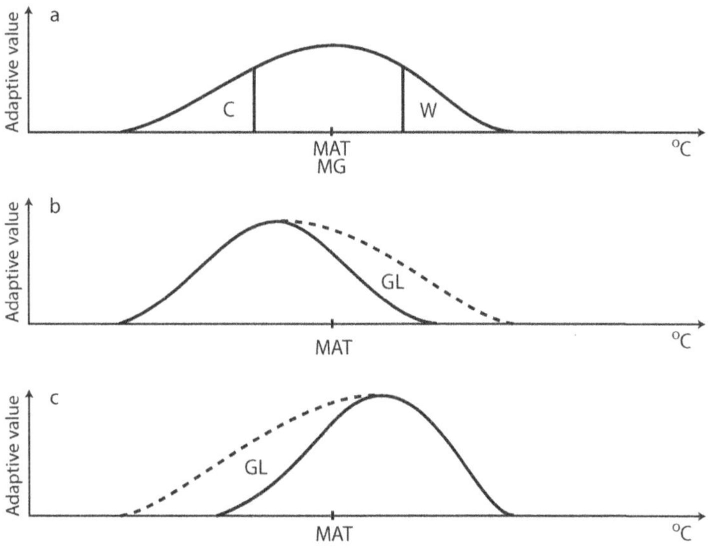

Fig. 10.1 Hypothetical year-to-year selection in young tree cohorts. Adaptive value here can be measured as the establishment probability of seedlings. Panel *a* shows the structure of a yearly seed cohort; panel *b*, the structure of the established seedling cohort in cold years; panel *c*, the structure of the established seedling cohort in warm years. *MAT* mean annual temperature, *MG* mean genotype, *C* cold year genotypes, *W* warm year genotypes. *GL* genetic load or maladapted genotypes. (From Lekevičius and Loreau (2012), the model is produced on the basis of data by Kelly et al. (2003))

natural selection shifts its direction not on seasonal, but on yearly basis. Therefore, this type of selection could be branded a 'year-to-year' selection.

It should only be added that annual climatic fluctuations could invoke population-level (not limited to cohorts) shifts of genotype frequencies only in such case if a year highly unfavourable to some genotypes completely eliminated them from the population.

Let us briefly focus on another question: can variations of the biotic environment in time also support genotype diversity? In the previous chapter, I wrote that different genotypes are best adapted to their 'own' biotic environment and are less successful in a 'foreign' one. However, this picture can hardly be considered complete since usually a biotic niche is not static. The environment which fluctuates around some intermediate values should support corresponding polymorphism, right? Is that really so? Among others, Pimentel with collaborators attempted to answer this question. At first, he presented the idea (Pimentel, 1961) that polymorphism should exist in wild populations as the response to the pressure from their exploiters: as the pressure of the consumers increases, the resistance of the exploited population should increase, since the most vulnerable genotypes are eliminated. And on the contrary, as the pressure of the consumers decreases, the most vulnerable genotypes would get an advantage and their frequencies would increase. The author branded this mechanism a 'genetic feedback' (Fig. 10.2). The part of the scheme which shows that decline in consumer population leads towards decreased resistance of the exploited population calls for a comment. Pimentel gives the following explanation: under those circumstances, the genotypes which are sensitive to the exploiters have

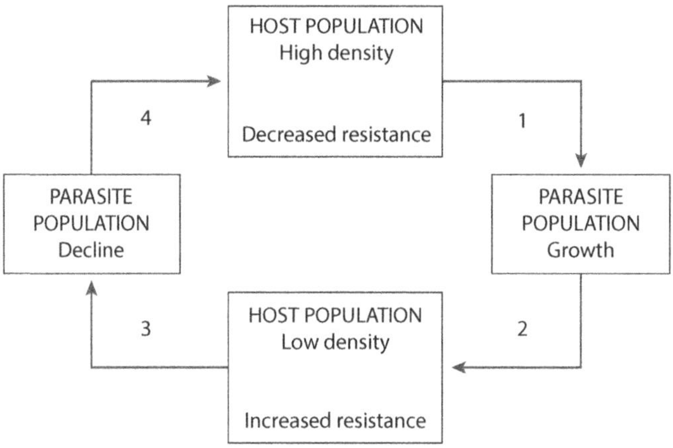

Fig. 10.2 The scheme of genetic feedback. In a dense host population, the genotypes that are vulnerable to parasites prevail (1); this provides favourable conditions for the increase of parasite population (2); the host population is reduced, the parasites eliminate the vulnerable genotypes promoting the dominance of the resistant ones (3); this reduces spreading opportunities for the parasites, suppresses their population, and the vulnerable host genotypes regain reproductive advantage (4). According to Pimentel (1968), such a mechanism of abundance regulation might be common to all sorts of systems: parasite–host, predator–prey, and herbivore-plant.

an advantage over the resistant ones because they do not waste assimilated energy to maintain the unnecessary resistance. Thus, they save more energy for growth and reproduction.

Attempts have been made to discover the above-mentioned feedback by way of experiment. Perhaps the most spoken of were the laboratory experiments with the housefly *Musca domestica* and the parasitoid wasp *Nasonia vitripennis* (Pimentel & Al-Hafidh, 1965; Pimentel & Stone, 1968). The authors discovered that the host and parasitoid populations that had never been in contact before initially have a dramatic impact on each other's abundance, so the fluctuations have a very high amplitude. Later, after a few dozen generations, the parasitoid reproduction rate and population size start declining while the host population recovers. Eventually the sizes of both populations settle down completely, although minor fluctuations persist. The ultimate result is moderate exploitation. When the flies that had gained certain level of resistance during the experiment were kept in parasitoid-free environment, their vulnerability to them was largely restored within 10 generations. This seemed to support the hypothesis set by the authors. They concluded that the fly population must have had genetic diversity in terms of parasitoid resistance before the experiment started. During coevolution observed in the experiment, natural selection was acting against the genotypes with low parasitoid resistance. Thus, polymorphism of the host species served as a kind of buffer which protected the population from a complete extinction, on the one hand, and from uncontrolled growth (with the help of parasites), on the other.

Perhaps even more illustrative case is the story of the European rabbit in Australia (Burnet & White, 1972). Back in 1859, 12 rabbit pairs were brought to Australia. Soon, they have multiplied to the extent that the hunters were no longer able to cope with them. This was caused, among other things, by the lack of natural enemies of these grazers. By 1950, the rabbit population has approached 1 billion. After numerous other measures failed, in 1950 Australians brought a rabbit myxomatosis virus from South America and released it into the wild. It caused a mass pandemic with enormous death toll. Only the resistant genotypes survived. However, in this case – and that makes it different from the one discussed above (Fig. 10.2) – both the resistance of rabbits has increased, and the virulence of the virus has declined. It is thought that the virulence was reduced by an *in situ* mutation which produced a new strain of the virus. Thus, about a dozen years later, the rabbit population has recovered, although it has never managed to reach its previous abundance. The rapid phase of coevolution was over, the system has attained a relative balance, and the exploitation has become moderate. Although this hardly means that the rabbit population no longer includes certain part of the vulnerable genotypes, just as the virus population – the aggressive ones.

The question remains open to what extent polymorphism of this type and genetic feedback are actually widespread in nature. This topic lost its popularity quite soon after being developed, it has even disappeared from the textbooks, which leaves me puzzled. Even more so, bearing in mind the fact that genotype diversity and moderate exploitation is not an exception but rather a law of nature.

Polymorphism is likely to be supported also by fluctuations in numbers: at high population densities, genotypes with higher intraspecific competitiveness should have an advantage, while the loss of population density would favour less competitive individuals. This conclusion was drawn after long-lasting observations and experiments on the common dandelion, Drummond's phlox, narrow-leaved plantain, as well as fruit flies and some small rodents (Sukatschev, 1928; Birch, 1955, 1960; Chitty, 1967; Krebs et al., 1973; Antonovics & Levin, 1980; Mueller & Ayala, 1981). Sometimes, just as in case of *K*- and *r*-species, more competitive genotypes are called *K*-genotypes, whereas those prevailing in sparser populations are called *r*-genotypes. Such polymorphism should perform as a buffer, similar to the previously described genotype diversity, which determines the relationship between the consumers and those being consumed. Unfortunately, it is not clear how widespread is this type of polymorphism.

Finally, some researchers hold an opinion that polymorphism determined by interspecific competition dynamics must also exist. Pimentel (1968) offered the following interpretation of the fact that competing species seldom displace one another completely. When one of the interacting species is numerous, it becomes subject to selection favouring genotypes tolerant to intraspecific competition. They, however, are less resistant to interspecific competition. This complicates the situation of that species, even more so, because selection simultaneously promotes interspecific competitiveness of the second species. Consequently, the second species starts to gradually outcompete the first one and eventually becomes dominant. Then the two species swap their roles. Thus, fluctuations of the population sizes of competing species must be accompanied by corresponding adjustments in their genetic structure that stabilise the interacting populations. The author has performed some experiments to check his hypothesis (Pimentel et al., 1965) and the results seem to support it.

To conclude, we can state with a degree of certainty that:

- Genotype diversity in wild populations is maintained by both abiotic and biotic factors. Qualitative or quantitative fluctuations of these factors in time constantly change the vector of selection pressure which presumably modifies the genetic structure of the population, and none of its genotypes is allowed to gain a permanent advantage. As already discussed in the previous chapter, genotype diversity is also supported by spatial patchiness of the environment. These elements might account for a significant part of the polymorphism observed in nature.
- Such an approach introduces a radical change in the previous one which was based on the concept that natural populations are dominated by one, the most efficient genotype called the 'wild type', while all the other genotypes are mutants with poor adaptive value, so they are doomed to extinction (Fig. 10.3);
- Genetic diversity is a kind of population memory. Selection must have saved information about different environmental conditions, which natural populations have encountered in the past, in the gene pool of these populations; therefore, no matter what environmental changes might happen (within the normal limits), there is always an adequate adaptive response of the gene pool. Polymorphism

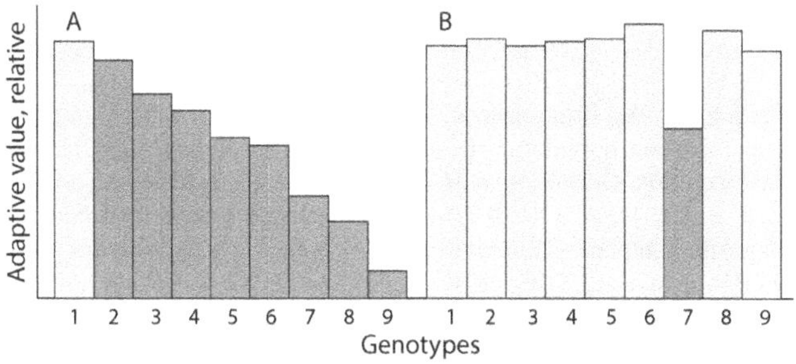

Fig. 10.3 Traditional (**A**) and new (**B**) approaches to genetic diversity in wild populations. The light columns represent 'wild' genotypes that are permanently present in the population; the dark columns stand for inadequate, temporary genotypes. (From Lekevičius, 1992)

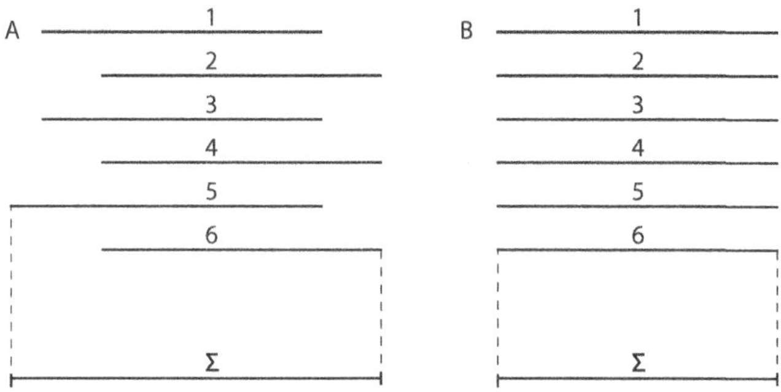

Fig. 10.4 Individual and population-level (the bottom line) reaction norms of two hypothetical populations: polymorphic (**A**) and monomorphic (**B**). For the purposes of this scheme, reaction norms could be interpreted as plasticity boundaries (for more detailed explanation, see Chap. 11). (From Lekevičius, 1992)

makes population viability, productivity, and biomass relatively independent from the whims of the environment. This is achieved by expanding the population-level reaction norms, compared to those which would be available to a population, monomorphic according to one of its genotypes (Fig. 10.4).

Note The term 'reaction norm' defines the limits, wherein phenotypic characteristics of an organism or a population can change in response to the changing environment. Since this variability of organisms is usually adaptive, the scope of the reaction norm also describes the adaptability or plasticity. To put it simple, there is a rather high probability that, regardless of the environmental changes within the usual limits, a heterogeneous population will include some individuals that will find the new conditions acceptable. They will start reproducing and will save the population from extinction. In a monomorphic population, however, an unfavourable environment will badly affect all individuals.

References

Ananina, G., Peixoto, A. A., Bitner-Mathé, B. C., Souza, W. N., Silva L. B. da, et al. (2004). Chromosomal inversion polymorphism in *Drosophila mediopunctata*: Seasonal, altitudinal, and latitudinal variation. *Genetics and Molecular Biology., 27*, 61–69.

Antonovics, J., & Levin, D. A. (1980). The ecological and genetic consequences of density-dependent regulation in plants. *Annual Review of Ecology and Systematics, 11*, 411–452.

Bergland, A. O., Behrman, E. L., O'Brien, K. R., Schmidt, P. S., & Petrov, D. A. (2014). Genomic evidence of rapid and stable adaptive oscillations over seasonal time scales in Drosophila. *PLoS Genetics., 10*(11), e1004775. https://doi.org/10.1371/journal.pgen.1004775

Birch, L. C. (1955). Selection in *Drosophila pseudoobscura* in relation to crowding. *Evolution, 9*, 389–399.

Birch, L. C. (1960). The genetic factor in population ecology. *American Naturalist., 94*, 5–24.

Burnet, F. M., White, D. O. 1972. Natural History of Infectious Disease. 4th Ed. Cambridge University Press.

Carvalho, G. R. (1987). The clonal ecology of *Daphnia magna* (Crustacea: Cladocera). II. Thermal differentiation among seasonal clones. *Journal of Animal Ecology., 56*, 469–478.

Chitty, D. (1967). The natural selection of self-regulatory behaviour in animal populations. *Proceedings of the Ecological Society of Australia., 2*, 51–78.

Dobzhansky, T. (1943). Genetics of natural populations. IX. Temporal changes in the composition of populations of *Drosophila pseudoobscura. Genetics, 28*, 162–186.

Dobzhansky, T. (1947). Adaptive changes induced by natural selection in wild populations of *Drosophila. Evolution, 1*, 1–16.

Dobzhansky, T., & Ayala, F. J. (1973). Temporal frequency changes of enzyme and chromosomal polymorphisms in natural populations of *Drosophila. Procceedings of the National Academy of Sciences USA., 70*, 680–683.

Ford, E. B. (1975). *Ecological Genetics* (4th ed.). Chapman and Hall.

Gurevičiūtė G., Lekevičius E., Kontrimavičius V. (Гурявичюте Г., Лекявичюс Э., Контримавичюс В.) (1989). The reaction of a polymorphic population of daphnia to seasonal temperature fluctuations. *Proceedings of the USSR Academy of Sciences, 306*, 250–252 (in Russian).

Kelly, C. K., Chase, M. W., De Bruijn, A., Fay, M. F., & Woodward, F. I. (2003). Temperature-based population segregation in birch. *Ecology Letters, 6*, 1–3.

King, C. E. (1972). Adaptation of rotifers to seasonal variation. *Ecology, 53*, 408–418.

Krebs, C. J., Gaines, M. S., Keller, B. L., Myers, J. H., & Tamarin, R. H. (1973). Population cycles in small rodents. *Science, 179*, 35–41.

Krimbas, C. B., & Powell, J. R. (1992). Introduction. In C. B. Krimbas & J. R. Powell (Eds.), *Drosophila inversion polymorphism* (pp. 1–52). CRC Press.

Lekevičius, E. (1992). Adaptation. In: E. J. Kormondy. *Ecological Concepts*. Littera Universitati Vytauti Magni. 253–309 (in Lithuanian).

Lekevičius, E. (1997). *A model of adaptive rearrangements on community, population, and organism levels. Habilitation work*. Institute of Ecology (in English and Lithuanian).

Lekevičius, E., & Loreau, M. (2012). Adaptability and functional stability in forest ecosystems: A hierarchical conceptual framework. *Ekologija (Vilnius)., 58*, 391–404.

Mueller, L. D., & Ayala, F. J. (1981). Trade-off between r-selection and K-selection in *Drosophila* populations. *Proceedings of the National Academy of Sciences USA., 78*, 1303–1305.

Mueller, L. D., Barr, L. G., & Ayala, F. J. (1985). Natural selection *vs*. random drift: Evidence from temporal variation in allele frequences in nature. *Genetics, 111*, 517–554.

Niklasson, M., Tomiuk, J., & Parker, E. D., Jr. (2004). Maintenance of clonal diversity in *Dipsa bifurcata* (Fallén, 1810) (Diptera: Lonchopteridae). I. Fluctuating seasonal selection moulds long-term coexistence. *Heredity, 93*, 62–71.

Pimentel, D. (1961). Animal population regulation by the genetic feed-back mechanism. *American Naturalist., 95*, 65–79.

Pimentel, D. (1968). Population regulation and genetic feedback. *Science, 159*, 1432–1437.
Pimentel, D., & Al-Hafidh, R. (1965). Ecological control of a parasite population by genetic evolution in the parasite-host system. *Annals of the Entomological Society of America., 58*, 1–6.
Pimentel, D., & Stone, F. A. (1968). Evolution and population ecology of parasite-host systems. *Canadian Entomologist., 100*, 655–662.
Pimentel, D., Feinberg, E. H., Wood, P. W., & Hayes, J. T. (1965). Selection, spatial distribution, and the coexistence of competing fly species. *American Naturalist., 99*, 97–109.
Rodriguez-Trelles, F., Alvarez, G., & Zapata, C. (1996). Time-series analysis of seasonal changes of the *O* inversion polymorphism of *Drosophila subobscura*. *Genetics, 142*, 179–187.
Rossi, V., & Menozzi, P. (1990). The clonal ecology of Heterocypris incongruens (Ostracoda). *Oikos, 57*, 388–398.
Sukatschev, V. N. (1928). Einige experimentelle Untersuchungen uber den Kampf ums Dasein zwischen Biotypen derselben Art. *Zeitschrift fur Induktive Abstammungs und Vererbungslehre., 47*, 54–74.
Weeks, A. R., & Hoffmann, A. A. (1998). Intense selection of mite clones in a heterogeneous environment. *Evolution, 52*, 1325–1333.

Chapter 11
Summing Up: Biodiversity and Plasticity of Life (General Adaptation Theory)

In the previous chapters, a lot of efforts were dedicated to present how species and genotype diversity helps communities and populations adapt to the ever-changing environment. It is high time to remember that the diversity of structures and functions existing within an organism plays a similar role. I mean the diversity of genes, macromolecules, metabolic pathways, and morphological and behavioural responses which define the phenomenon called phenotypic plasticity. Some authors add that these are rearrangements within the limits of individual reaction norms.

Let us focus on these adaptation mechanisms a little longer, because I suspect that not necessarily all my readers are well familiarised with them. Hochachka and Somero (2002; see also Somero et al., 2017) distinguish the following types of biochemical and physiological responses:

1. Cellular concentrations of enzymes already present before the environmental change are adjusted (quantitative strategy).
2. Production of new enzymes previously absent from the cells starts. They better match the needs of the organism under the new conditions (qualitative strategy). This involves no mutations and appearance of new genetic information but relies on the use of the existing information through transcription and translation.
3. Modification of the microenvironment of enzymes. For example, direct environmental impact is often reflected by changes of the composition and concentrations of dissolved low-molecular-weight compounds and ions within a cell. This in turn changes (usually with some adaptive benefits) the spatial structure and functions of macromolecules. Admittedly, this adaptation mechanism is poorly understood, yet it is not unusual or insignificant in nature.
4. An overall metabolic change, when alternative metabolic pathways are triggered. In this case, the rearrangements are essential, involving all or most organ systems. They can be triggered by both severe changes of the environment (e.g. early frosts) and various internal physiological signals enabling the organism to preadapt (e.g. transition from one phase of individual development into another or different endogenous rhythms). These external or internal signals can intensify or slow down the entire metabolism or substitute certain metabolic processes with different ones. Such response resulting in essential rearrangement of the organism's functions usually involves its regulatory apparatus – the neural and/or endocrine systems.

E. Lekevičius, *Biodiversity*, https://doi.org/10.1007/978-3-031-11582-0_11

There is no clear boundary between these four types of responses. Moreover, one should bear in mind that often several mechanisms get triggered simultaneously in a single response. More detailed data about physiological and biochemical mechanisms of adaptation to the changing environment could be found in Hochachka and Somero (2002), as well as some other overviews such as Pigliucci (2001), Larcher (2003), Willmer et al. (2005), Nicotra et al. (2010), and Krauss and Nies (2014).

Facts suggest that the threshold of organisms' sensitivity to environmental changes is very low or even non-existent and they are capable to respond even to minor and quite ordinary changes. For instance, Harris (1973) has discovered that within just 1–2 h after the sun is obscured by clouds biochemical characteristics of diatoms change, they transform from photophilic into shade-tolerant organisms. This allows diatoms to adapt to the current cloud cover almost every hour and to maintain rather high levels of photosynthesis even on gloomy days. Buchner and Neuner (2001) discovered that the highland plant *Silene acaulis* takes about 3 h to adapt to the warming weather (usually every noon). Thus, organisms try to tune up their physiological optimum to the changing environmental parameters. The above-mentioned biochemical mechanisms play a key role also in these rearrangements.

To experience our own adaptation mechanisms in action, it would suffice to go to the mountains and to reach the altitude of some 3–4 thousand metres. The first few days, one would feel really bad, experiencing lack of oxygen, intensified heartbeat, increased perspiration, sluggishness, and apathy; even reasoning and speech would be affected. One would draw the words out of their mouth, as if forced to speak. Within 2–3 weeks, however, the organism would adapt. Haemoglobin concentration in blood would have increased, the heartbeat would have normalised, the breathing would have become slower and deeper, while the speech – more coherent. On descent one would have to adapt again, this time to the usual conditions. Haemoglobin content and other physiological parameters would revert to their 'normal' values within a few days.

Young (2002), an expert in circadian rhythms, writes:

> Animal behaviour is cyclical, with regular patterns of activity and inactivity alternating in synchrony with environmental cycles of day and night. Associated with these behavioural rhythms are robust physiological oscillations of enzyme activity, hormone synthesis and release, body temperature (for mammals) and gene expression (...) These 'circadian' rhythms are not restricted to the animal kingdom. Some of the earliest experimental evidence for endogenous circadian clocks came from studies of plants and fungi, and recent work in photosynthetic bacteria has demonstrated that in some species most genes switch on and off with a 24-h regularity imposed by a circadian clock.

According to the author, all these cyclical rearrangements have an adaptive character. One could easily find numerous similar examples in the living world, and they are well known to experts. Organisms are rather well pre-adapted to daily, or circadian, phenomena – the alternation of light and darkness (photoperiodism) and the associated temperature changes. Responses to them involve changes in gene and enzyme activity as well as hormone production that plants and animals initiate automatically prior to the actual change of the environment. Adaptations to seasonal rhythms happen in a similar manner. However, living systems are always somewhat late to respond to directional and random changes. This is understandable, since adaptive rearrangements take time; they are not instant. Therefore, we would hardly step away from the truth by stating that organisms are constantly adapting to their environment, and it is a never-ending process: they hardly finish adapting to one change, when another is already taking place, and that goes on and on...

Experts in ecological physiology are inclined to interpret the process of adaptation as metabolic rearrangements that enable an organism to optimise its functions, so that they better match the changing environment (Rosen, 1967; Somero et al., 2017). I hope there is enough evidence (see Chaps. 7 and 10) that adaptive rearrangements are also possible at the levels of a guild or population, where they are likewise related with the optima shift in one direction or another. Thus, it seems appropriate to assert that adaptation to the changing conditions is a universal biological phenomenon constantly occurring at all organisational levels. Let us try to define this aspect of adaptation more strictly by using a graphic model (Fig. 11.1).

Note One should distinguish between a direct environmental impact (arrow 1) and the response to the environmental change (arrow 2). A direct impact is observed immediately after an abrupt change of the environment, prior to the adaptation. Thus, the values are presented as a curve which could be called the curve of tolerance. On the other hand, when the environmental change is slow, or when biological activity is measured long time after the change, we get a curve of plasticity or adaptability instead of that of tolerance. Therefore, the number of the curves of tolerance is infinite, and they may slide both ways along the environmental gradient. The curve of plasticity, however, is always broader than any of the curves of tolerance, and its shifts are possible only as the result of evolution. A similar situation is with the optima. A single organism may have an infinite number of the tolerance curve peaks whose projections against the x-axis could be called optima. That does not mean, however, that all of them are equal. As seen from the picture, the curves of tolerance not only shift along the axis, but their height can differ as well; therefore, the ideal homeostatic plateau often presented in textbooks is seldom completely

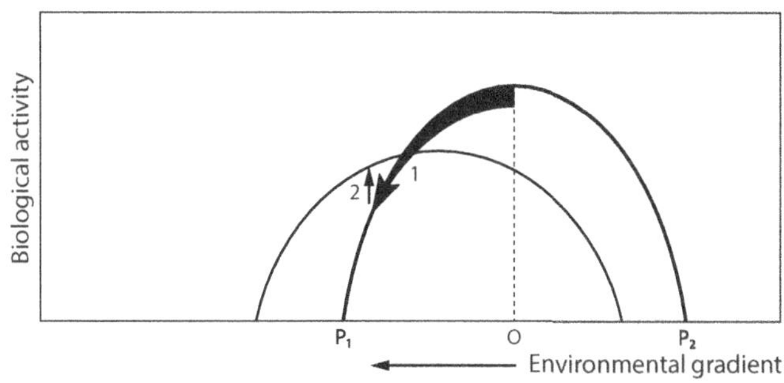

Fig. 11.1 The response of organisms to an environmental gradient may be represented by a shift of tolerance curves. Here O is an optimum and P_1 and P_2 are two pessima. Biological activity or adaptedness decreases (1) following an abrupt worsening of environmental conditions (Δ), but it may be partly or wholly restored after some time due to adaptation (2)

achieved in reality, although the trend does exist. The environmental conditions promoting the maximum biological activity among all the possible options could be labelled the main optimum. All other observed optima should be considered as auxiliary, a sort of 'emergency measures'.

The process of optimum adjustment can be observed directly. One of the most famous options is the thermogradient method. It is based on a device whose main component is a long tunnel filled with water. One end of this tunnel is heated, while the other is cooled down; so eventually a temperature gradient ranging from, say, 0 °C at one end to +35 °C at the other is produced. Then the experimental animals, for example, fish or crustaceans, are released into the tunnel. The core of the method is the ability of organisms to make an unrestricted choice of the preferred temperature within the given range. Experts are convinced that they choose the temperature that best matches their current physiological state. Thus, by registering the positions of the organisms in the thermogradient tunnel, the researcher can also see the shifts of the optimum points during the process of adaptation. Naturally, the speed of shifting of these points can be interpreted as the rate of adaptation (Fig. 11.2).

Now we could move on to the theoretical synthesis. Ashby (1956), one of the creators of cybernetics, once expressed an idea which he named the law of requisite variety. According to Ashby, from the functional point of view, an organism is a set of essential and subsidiary variables. Examples of essential variables could be brain oxygen level, blood glucose content, or body temperature of a mammal. These are the key parameters, and the state or even survival of the organism depends on its ability to stabilise them (to reach homeostasis). Conversely, subsidiary variables may and should vary within broad limits, since it is this variation that stabilises the essential variables regardless of the uncertainty of environmental conditions.

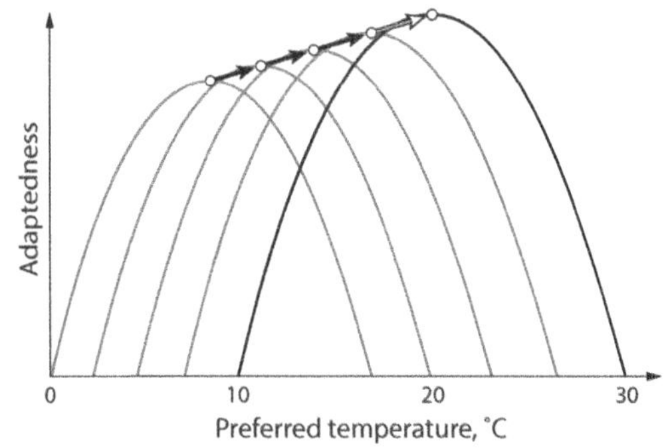

Fig. 11.2 Organisms released into a thermogradient tunnel usually gather in a particular area of the tunnel and then slowly move towards their most preferred temperature (the main optimum). This movement is thought to indicate the shift of the optimum points (the arrows connecting the peaks of the curves) and to represent the processes of physiological adaptation taking place in the organisms. The graph represents behaviour of the organisms sampled from a water body in late autumn. (From Lekevičius (1992), the model is produced on the basis of experimental data by Juknevičius and Lekevičius (1991))

Table 11.1 Adaptation mechanisms observed in living nature

Adaptation mechanisms	Short characteristics
Individual	Differential expression of genes (phenotypic plasticity)
Populational	Differential survival and reproduction of individuals (natural selection)
Biocenotic (community-level)	Differential changes of population size (ecological succession)
Genetic or evolutionary	Genetic variation and natural selection

From Lekevičius (1986), see also Lekevičius (2011)

Fluctuations of these variables are responses to the environmental changes, and they usually have an adaptive nature. The following scheme applies to living systems:

$$D \rightarrow R \rightarrow E$$

where D represents variation of the environmental conditions, R is a regulator, and E is a set of essential variables. R acts to minimise the impact of D on E. The necessity to secure stability of the essential variables requires that an organism be equipped with the diversity of responses at least as rich as the diversity of environmental conditions. In other words, homeostasis is only possible if the internal variety (of responses) is capable of fully neutralising the external variety. And this is exactly the 'requisite variety'.

M. Conrad was the first to expand this concept and to apply it to all adaptation phenomena. This author began to develop his hierarchical adaptability theory (HAT) over 40 years ago and has published a few dozens of articles on this topic. Later he summarised his works in a monograph (Conrad, 1983). He made use of some equations from information theory. The other part of the mathematical apparatus was his own creation. Conrad's main idea was as follows: not only individual organisms but also populations and ecological communities are able to bring their essential variables into homeostasis. Behavioural diversity of a community consists of responses which belong to it as a whole, as well as to all the lower levels of organisational hierarchy down to single cells and macromolecules. The higher functional and structural diversity, the broader the range of potential responses. Following the tradition of information theory, Conrad refers to the latter as entropy, or community's behavioural uncertainty. Likewise, the diversity of environmental conditions could be interpreted as environmental uncertainty.

Soon after Conrad published his monograph, another similar concept was presented in the academic press under the title of General Adaptation Theory (GAT, Lekevičius, 1986; see also Lekevičius, 2011). It differed from Conrad's version first of all in terms of methodology – this time, hierarchical systems theory, causal analysis, and conceptual modelling were put into play. In addition to these authors, a few more researchers of the last decades of the previous century could be mentioned, whose works might be considered as rather independent calls for a single model comprising all the information about the ways of adaptation known to science. First of all, one should mention Shkorbatov (1971, 1982), Shilov (1988), and a handful of other scholars who had significantly contributed to the development of this theory (see the reviews in Lekevičius, 2007, 2011). All these researchers shared a common idea: diversity of genes, genotypes, and species is the sole instrument at the disposal of life for a successful neutralisation of environmental variation. Hereafter, I am going to use mostly the postulates of GAT, although some references will also be made to Conrad's conclusions.

Ashby's statement that a certain minimum level of biodiversity is necessary to make adaptation and homeostasis possible should be interpreted as follows in the context of HAT and GAT: biodiversity helps organisms, populations, and communities to adjust their corresponding optima to the changing environment. To sum up what has already been presented earlier (Chaps. 7 and 10), one could distinguish the following mechanisms of adaptation:

Note The first three mechanisms could be called non-genetic in the sense that adaptation in these cases does not require the occurrence of new genetic information by way of mutations or recombination. Therefore, they can also be categorised as reversible, whereas evolution is usually considered an irreversible process. When talking about biocenotic mechanisms of adaptation (differential changes of population size = replacement of the dominant species = ecological succession), we usually bear in mind the cases when species of the same guild substitute or compensate for each other. This is only natural, since organisms representing different guilds cannot substitute one another. Neither can organisms from different trophic levels or functional kingdoms. Consequently, the stability of trophic levels and functional kingdoms is secured through the plasticity of their constituent guilds, and whenever this plasticity is exhausted, the entire community will face catastrophic consequences of various scale

Mechanisms of adaptation are subject to hierarchy comparable to that of corresponding organisational levels. This means that, for instance, when a seasonal succession of phytoplankton takes place, a seasonal selection (differential reproduction of pre-adapted genotypes) should also be happening along with the rearrangements within the scope of individual reaction norms. If this is the case – and the evidence seems to be sufficient – then rearrangements at individual and population levels might be considered as the tools for a more delicate adjustment of the entire community's optimum to the environmental conditions and as a series of defence lines along the way of a stressful impact towards the essential variables of the ecosystem. These variables are supposed to be the most resistant to destabilisation.

Adaptation mechanisms differ significantly in terms of their rates. Individual rearrangements are relatively fast, especially the behavioural and biochemical responses. Physiological and morphological rearrangements last much longer, sometimes up to a few weeks. The rate of individual responses is mostly limited by the speed of neural impulses, synthesis of materials, and mitosis. The reader must have realised by now that adaptive rearrangements at population and community levels are even slower since their rate is limited by fecundity and generation time. Nonetheless, these rearrangements are much faster than evolution since, in addition to the already mentioned factors, its rate is also limited by mutation and recombination rate.

Another component of adaptability, or plasticity, is plasticity boundaries. They depend on the range of responses or, more precisely, the diversity of biological structures available within that or another mechanism. Strictly speaking, plasticity boundaries represent the displacement range of the tolerance curve along the axis of environmental conditions, while adaptation rate reflects the displacement speed of the curve (see Figs. 11.2 and 11.3).

So, ranking of the mechanisms according to the plasticity boundaries is directly opposite to their order according to the adaptation rate: individual rearrangements allow for adaptation within a relatively narrow interval of environmental conditions, evolution has practically unlimited potential, whereas populational and community-level mechanisms fit somewhere in between (Fig. 11.3).

Theoreticians should ask themselves: why are adaptation mechanisms designed in that particular way? Why did life need such a variety of adaptation forms? One possible answer, albeit rather general and possibly not very precise, could be as follows: because organisms face very different changes in their environment – random and periodic, negligible and dramatic, rapid and gradual, and usual and unexpected; so where one form of adaptation fails, the other might be successful. An individual must be active here and now, it cannot afford waiting a month or a year for its metabolism to adjust – it will be too late. Besides, it is engaged in a competitive struggle with other individuals which might recover sooner. Meanwhile, a population can wait, especially, if it managed to preserve most of its gene pool and thus is able to reconstruct itself. Take a unicellular organism – its generation time is so short that its environment remains practically stable, and there is no need for a broad reaction norm. When the conditions change, it will die easily but will be successfully replaced by other individuals of the same species with a different genotype. Death of poorly adapted or inefficient individuals is one of life's greatest inventions, since it makes possible to decompose the detritus and use it for generating new individuals that are better adapted to the altered conditions. As the wheel of birth and death keeps turning, populations become almost immortal. An ecological community is in no hurry; it is not involved in any competition; it is focused on plasticity

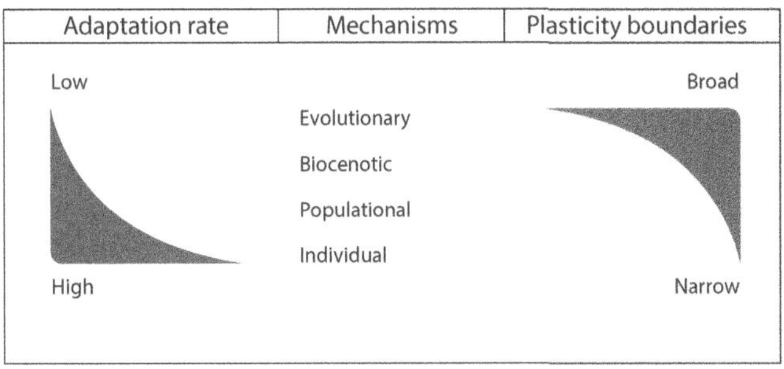

Fig. 11.3 Characteristics of adaptation mechanisms taking into consideration both components of plasticity – adaptation rate and plasticity boundaries. (From Lekevičius, 2011)

boundaries, rather than speed; and it can recover even after it has lost a significant number of species.

It was mentioned earlier that rearrangements within the boundaries of individual reaction norms take place without changing the genotype. Simply some silent loci are activated in an appropriate moment or place, and their transcription starts, or the performance of already working genes is intensified through increased translation (qualitative and quantitative strategies – Hochachka & Somero, 2002). As we could hopefully learn from the material presented in the previous chapters, analogous processes take place at population and community levels: a usual change of the environment gives a reproductive advantage to certain genotypes or species and supresses the activities of some other genotypes and species. It appears that any elementary (molecular) function can be triggered and strengthened in the following ways:

- By initiating transcription and translation of a formerly 'silent' genetic locus which has recently become desirable (qualitative biochemical strategy).
- By increasing the concentrations of a produced molecule in a cell (quantitative biochemical strategy).
- By way of mitosis of these cells.
- By increasing the frequencies of the corresponding gene within the population.
- Through the growth of the entire population.

Thus, the number of copies of a particular molecule can reach astronomic heights and become so significant that it can push the tolerance curve of the entire population or even ecological community in the right direction.

In fact, non-genetic mechanisms of adaptation, ranging from individual to biocenotic, can be seen as a kind of amplification chains of enzyme functions. Should we not focus exclusively on non-genetic methods of adaptation, then the sixth option could be added to the list above: a possible split of a metapopulation which carries a new genetic information into separate populations eventually leading to the occurrence of new species. All these mechanisms could be called the methods of strengthening of genetic information. Perhaps the first to present a similar, albeit somewhat simplified and incomplete, vision of the path of a successful mutation from its very occurrence was the Russian evolutionary biologist N.W. Timofeeff-Ressovsky. I should only add that, when certain information is amplified, there is another set of information which is suppressed because it is no longer supported by the environment: macromolecules degrade; active loci become silenced; and organisms cease reproducing, may enter a cryptobiotic state, are eliminated by the competitors, or produce dormant propagules – dormant seeds, spores, buds, tubers, eggs, or cysts. This information represents 'memories' of the past events, stored in the genetic material, which can be strengthened again and used for the adjustment of optima when the right moment comes (sometimes many years later).

Let us analyse an example to have a clearer picture of real interactions between the mechanisms of strengthening/suppression of elementary functions. A guild of forest trees perfectly adapts to daily and seasonal fluctuations by using only the

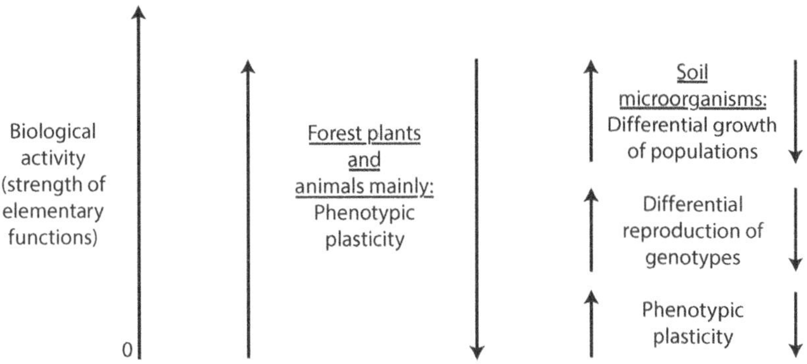

Fig. 11.4 Likely responses of the main components of a temperate forest ecosystem to diurnal, day-to-day, and seasonal fluctuations in abiotic conditions. Elementary functions are enzymatic functions, and their reinforcement is due to biosynthesis of macromolecules, mitosis, and reproduction of organisms. (From Lekevičius & Loreau, 2012)

diverse macromolecules and metabolic pathways, together with morphological responses, i.e. only phenotypic plasticity (Fig. 11.4).

Forest animals, whose generation time lasts for months or years, represent a similar case. However, the guild of soil microorganisms (decomposers) is likely to achieve the same result by also involving the changes in frequencies of pre-adapted genotypes as well as the changes of dominant species. The fact of successful coexistence probably indicates that the guild of microorganisms is equipped with a similar level of adaptability (adaptation rate and plasticity boundaries) to that of an *individual* tree, although it uses additional mechanisms of strengthening and suppression of functions to achieve that adjustment.

The picture above presents a situation which illustrates quite well the principle of compensation of adaptation mechanisms formulated by Conrad (1983). If certain organisms have highly developed individual mechanisms of adaptation, one should expect, according to Conrad, that they are inferior to other species in terms of other components of adaptability, e.g. ability to adapt through populational and evolutionary mechanisms. For example, humans and other advanced mammals have low evolvability and undeveloped populational mechanisms, but they use effective behavioural and immune responses.

To sum up, everything that we have learned about the role of biodiversity within the last few decades, one may clearly state the following. Diversity is useful to life, first of all as a tool of specialisation with regard to the biotic and abiotic environment. Moreover, it is also an instrument of adaptation to the environmental fluctuations in time and space. As the environment changes, certain structures are activated, while others performing analogous functions lose their optima and are suppressed. Therefore, at any particular time and in any particular location, only a part of available structures and functions are put to use, while the others should be considered as a reserve. They are dormant and waiting for the right moment, and usually that moment arrives sooner or later. Therefore, no matter what changes take place in the environment within its normal scope, there is always a proper structure and an adequate response in the overall set of responses, and that makes life a system which never deviates too far from the optimum and sometimes even attains it. In short, it would be a grave mistake to consider the dormant loci, rare or not very active macromolecules, or exotic genotypes and species as redundant structures. A much more appropriate term would be 'genetic memory' or a similar expression which more adequately reflects the real state of affairs.

The statements applicable to local communities should, to a larger or smaller extent, apply also to their entirety, the biosphere. It should also be capable to bring its functions into homeostasis by using not only the diversity of genes, genotypes, and species but also diversity of communities. In my understanding, this is exactly what the Gaia hypothesis is all about (Lovelock & Margulis, 1974; Margulis & Lovelock, 1974; Watson & Lovelock, 1983). The authors of this hypothesis have presented numerous ideas, some of them inconsistent or even contradicting each other. No wonder then that the authors who addressed it offer different interpretations. My own version might also appear a bit unconventional, but I hope it does not distort the essence. In my understanding, the hypothesis can be described by the following statements:

- The present state of the atmosphere, hydrosphere, and lithosphere has been significantly shaped by the activities of life.
- In the course of evolution, the impact of life on the abiotic environment has been steadily increasing. Acting randomly, life has been modifying the properties of the atmosphere, hydrosphere, and lithosphere, often to its own detriment (the changes were a by-product of these activities). However, in the long run, life would adapt to the new environment, so that it would no longer be hostile but, on the contrary, favourable, or even optimal.
- To some extent, life is capable of maintaining an optimal abiotic environment thanks to differential activity of species and guilds. Thus, the properties of the atmosphere, hydrosphere, and lithosphere are emergent, since neither a single species of the biosphere nor even an entire guild are able to bring them into homeostasis. The biosphere is an organised system capable of self-arrangement.

The exceptional feature of the Gaia hypothesis is that it foresees a possibility to stabilise and optimise the non-living environment. This possibility is realised thanks to species diversity. I consider this hypothesis (though 'theory' would be a more appropriate term) an extension of GAT and HAT and their extrapolation to the biosphere level. But it would be equally correct to consider GAT and HAT as supplements to the Gaia hypothesis. The authors of the hypothesis themselves might not share this opinion; we do not know that. In any case, it is very unlikely that they were aware of M. Condrad's and my works. Perhaps I should also mention some of my own publications on the topic of coevolution of the non-living spheres of our planet and its biota (e.g. Lekevičius, 2000, 2002, 2013; see also Chap. 18 herein). My vision of this coevolution has not been inspired by the Gaia hypothesis; it came from other sources. The conclusions I came up with are not in conflict with any of the three statements listed above.

Last but not the least, one more important question remains open: what is the direction of life's never-ending efforts? The shifting of tolerance curves should serve some purpose. In other words, which indicators are optimised and what are these 'essential variables' (W. R. Ashby) on the scale of the entire biosphere as well as of individual communities and populations. I referred to them as biological activity and adaptedness (Figs. 11.1 and 11.2), but I did not explain what I meant by these terms.

To begin with, understanding of the functional and control hierarchies (Chap. 3) and of the hierarchy of adaptation mechanisms (this chapter) leads to a quite logical conclusion, as I see it, that the 'goal functions' characteristic to the components of an ecological community, ranging from an individual up to the community itself, should not be very different. This difference is no bigger than the difference between, say, the goals of an individual entrepreneur and the goals of country's economy – the country cannot get wealthy if its citizens do not share this goal; however, these citizens might only be interested in their personal wealth. Nature is no different. An equivalent of wealth in nature would most likely be the biomass, since it is the biomass that organisms try to maintain stable, regardless of the unstable environment, and to effectively increase it at the slightest occasion. If we were to speak only about the organisms, populations, and communities in their stationary phases, maintenance of the biomass on a stable and the highest possible level can only be achieved by way of upholding and increasing their functional efficiency (Fig. 11.5). It can be defined as the highest possible proportion of the final product, which is biomass in this case, in comparison with the total amount of resources used to produce it. In short, nature cannot afford to be wasteful in the world where resources are almost

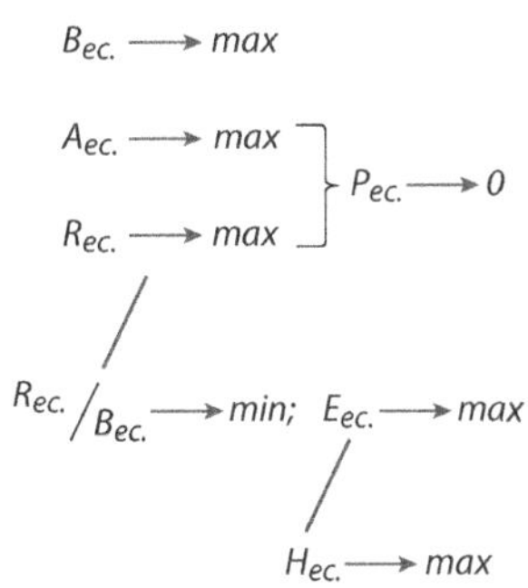

Fig. 11.5 Hypothetic relations between the main ecosystem-level indicators (*ec.*) which are likely to be optimised with the help of all available adaptation mechanisms. The top part presents the four main indicators: the biomass of the ecosystem ($B_{ec.}$), its assimilation ($A_{ec.}$), respiration ($R_{ec.}$), and production ($P_{ec.}$). The trends presented herein are typical of an ecosystem which is close to the stationary state ($P_{ec.} = 0$; $B_{ec.} = const.$). They are determined first of all by the growth of functional efficiency ($E_{ec.}$) linked with the reduction of costs to support biomass viability ($R_{ec.}/B_{ec.}$) and increased specialisation level which presumably has a positive correlation with ($H_{ec.}$), species diversity. (From Lekevičius (1986), with modifications)

always sparse, so it strives for efficiency, which is the highest when material and energy losses are minimised.

Such an approach is quite in tune with the well-known-in-physics Prigogine's theorem, which states that as a system approaches its stationary state, the rate of entropy production is minimised, i.e. the system uses less and less free energy for self-maintenance. In my understanding, this idea is also reflected in the optimality principle once formulated by Rosen (1967), one of the most famous theoretical biologists of the twentieth century. By the way, the classics of ecology, Odum (1969), Whittaker (1975), and Hannon (1979) also considered that, despite various disturbances, ecological communities 'strive' in all possible ways to minimise energy consumption for keeping a biomass unit viable and thus increase total biomass to the maximum possible level under the given circumstances. According to Veldhuis and co-authors (Veldhuis et al., 2018), ubiquitous and constantly functioning 'within-cycle competition (between different plants or decomposers) is a force that spontaneously leads to more materially closed and more productive ecosystems'. Leigh and Vermeij (Leigh Jr & Vermeij, 2002; Vermeij, 2005) present a similar description of life's 'strategic goals'. In their opinion, throughout evolution, species diversity was growing almost continuously, and this process was accompanied by the increase of specialisation or labour division levels which, in turn, ensured more efficient material cycles and larger productivity of ecosystems.

Although this problem is not an easy one, the objectives of this monograph are such that we should not focus too much on that topic; moreover, that has been addressed in detail in my other publications (e.g. Lekevičius, 1986, 2011). Therefore, I am going to stick to the idea that adjustment of communities to the environment means that they attain maximal possible total biomass. Populations (and species) also 'strive' towards the same goal, although each of them, obviously, increases only its own biomass, often at the expense of other species.

Now we could try to answer the question raised in the Preface of this monograph: could life do without the diversity of individuals and species; could there be different types of life on Earth, in the form of a continuous film covering the entire surface of the planet? In such a case, there would be neither multitude of individuals nor species, as we understand them – instead, there would be a vast community stretching all over the place as a single unit, a superorganism in its true sense. Perhaps such an organism would have its internal waste-free material cycle, being both a producer and a decomposer at the same time because it would hardly last long otherwise.

What capabilities to survive and adapt would such a living being have? The first thing to come to mind when seeking for an answer is this: such a life form would have to rely only on individual adaptation mechanisms; it would not have any other mechanisms that would facilitate its day-to-day functioning. Secondly, if the environmental conditions fluctuated the way they do on our planet, that superorganism would have to possess an unimaginably wide reaction norm and plasticity boundaries.

But even it that case, such variation of environmental conditions in time as we usually observe on Earth would most likely be a real challenge for the superorganism. For example, it is difficult to imagine this enormous construction starting to move and migrate in order to remain in the optimum zone as the climate warms up. Meanwhile, our individuals and species – even trees! (in the form of seeds) – seem to be capable of migrating and have already done than more than once in historical past.

Another problem is also likely to arise in connection with the control of activities of such a giant. In my understanding, a superorganism like that, if it could ever evolve at all, could only be controlled in a centralized way, just like a planned economy or a multicellular organism. Perhaps such a superorganism would be controlled by a 'superbrain', or maybe it would have a different solution, but in any case, any simple change of the environment could complicate the situation just because its parts would not be independent and would wait for the instructions 'from above'. It is difficult to perceive the volume of information that such brain would have to process.

Even more important is the fact that maximum level of adaptedness is incompatible with equally high adaptability (Lekevičius, 1986). An increase of specialised organisms within a community has a positive effect on its functional efficiency (see also Schoener, 1971) but only to the extent when it starts interfering with the community's ability to bring global parameters into homeostasis. This may be the reason why in wild nature, along with numerous specialised organisms, we find quite many generalists, such as euryphages and eurytherms. Generalists are likely to not only weaken the integration but also add more plasticity to the entire system. We find a similar conclusion in HAT: the stockpiled reserve or 'redundant' functions and structures as well as weak interactions increase plasticity of biological systems but decrease their functional efficiency (Conrad, 1990; see also McCann et al., 1998). Conrad calls this regularity 'multiple weak interactions'.

Nevertheless, evolution has produced a multitude of strongly integrated structures which we call organisms or individuals. In this way it has been 'striving' for efficiency. However, by not creating a superorganism and granting no immortality to individuals, it rendered them vulnerable and made them ephemeral residents in transient environment. This may be the reason why short-living unicellular organisms have narrower individual reaction norms than the multicellular ones with much longer life spans. Admittedly, polymorphism and ability to evolve fast seem to compensate for the lack of phenotype plasticity in unicellular organisms. Thus, their populational plasticity is more or less comparable to that of multicellular life forms. All this suggests that our biosphere has been 'striving' to achieve – and indeed has achieved – a certain balance or a compromise between functional efficiency and plasticity. Then again, *Homo sapiens* seems to have violated this balance: its population has a reaction norm that significantly exceeds those of other multicellular life forms, while the functional efficiency of its organism is hardly inferior to that of other large mammals.

To sum up, it is difficult to imagine the occurrence of a single superorganism, even in theory. Perhaps only on a planet where the climate is much more stable than that on Earth. It makes little sense to further speculate about the possible characteristics of such a superorganism, since we have never encountered anything like that, and it is very easy to let imagination run wild. In my opinion, one thing is obvious though: a superorganism would be very efficient in terms of functioning and could generate a total biomass exceeding that of our biosphere but only under very constant conditions.

Conversely, on a planet like our Earth where environmental conditions are ever changing, life might have only been able to survive by distributing its functions to structures separated in space and enjoying certain degree of independence. Management of 'the economy of nature' was destined to – and indeed has – become decentralised, so that the right of initiative belongs to species and individuals. Just like in a wild capitalism (Lekevičius, 2009).

References

Ashby, W. R. (1956). *An introduction to cybernetics*. Wiley.

Buchner, O., & Neuner, G. (2001). Determination of foliar heat tolerance: A new equipment for field measurements. *Journal of Applied Botany, 75*, 130–137.

Conrad, M. (1983). *Adaptability: The significance of variability from molecule to ecosystem*. Plenum Press.

Conrad, M. (1990). The geometry of evolution. *Bio Systems, 24*, 61–81.

Hannon, B. (1979). Total energy costs in ecosystems. *Journal of Theoretical Biology, 80*, 271–293.

Harris, G. P. (1973). Diurnal and annual cycles of net plankton photosynthesis in Lake Ontario. *Journal of the Fisheries Research Board of Canada, 30*, 1779–1787.

Hochachka, P. W., & Somero, G. N. (2002). *Biochemical adaptation*. Oxford University Press.

Juknevičius V., Lekevičius E. (Юкнявичюс В., Лекявичюс Э.) (1991). Thermal preference of young crayfish *Pacifastacus leniusculus* (Dana) during summer-autumn period. *Ekologija (Vilnius), 3*, 11–17. (in Russian).

Krauss, G.-J., & Nies, D. H. (Eds.). (2014). *Ecological biochemistry: Environmental and interspecies interactions*. Wiley – VCH Verlag.

Larcher, W. (2003). *Physiological plant ecology* (4th ed.). Springer.

Leigh, E. G., Jr., & Vermeij, G. J. (2002). Does natural selection organize ecosystems for the maintenance of high productivity and diversity? *Philosophical Transactions of the Royal Society of London B, 357*, 709–718.

Lekevičius E. (Лекявичюс Э.) (1986). *Elements of a general adaptation theory*. Mokslas Publ. (in Russsian).

Lekevičius, E. (1992). Adaptation. In E. J. Kormondy (Ed.), *Ecological concepts* (pp. 253–309). Littera Universitati Vytauti Magni. (in Lithuanian).

Lekevičius, E. (2000). *Only an ecosystem is alive: A somewhat nontraditional approach to the evolution of life*. Vilnius University Publ. (in Lithuanian).

Lekevičius, E. (2002). *The origin of ecosystems by means of natural selection*. Institute of Ecology Publ.

Lekevičius, E. (2007). Adaptability of life and its ability to adapt to anthropogenic changes (a theoretical study). In M. Žalakevičius, V. Būda, & E. Lekevičius (Eds.), *Biota and global changes* (Vol. 1, pp. 8–78). Institute of Ecology Publ. (in Lithuanian).

Lekevičius E. (Лекявичюс Э.). (2009). On some analogies between ecosystem evolution and development of economy: From A. Smith and C. Darwin to the latest ideas. In L. E. Grinin, A. V. Markov, & A. V. Korotaev (Eds.), *Evolution: Cosmic, biological, and social* (pp. 226–260). Librokom. (in Russian).

Lekevičius, E. (2011). A review of and some comments on the attempts to construct an overarching theory of ecosystems plasticity. *Ekologija (Vilnius), 57*, 137–153.

Lekevičius, E. (2013). *Ecology: From individuals to the biosphere*. Vilnius University Publ. (in Lithuanian).
Lekevičius, E., & Loreau, M. (2012). Adaptability and functional stability in forest ecosystems: A hierarchical conceptual framework. *Ekologija (Vilnius), 58*, 391–404.
Lovelock, J. E., & Margulis, L. (1974). Atmospheric homeostasis by and for the biosphere - the Gaia hypothesis. *Tellus, 26*, 2–10.
Margulis, L., & Lovelock, J. E. (1974). Biological modulation of the earth's atmosphere. *Icarus, 21*, 471–489.
McCann, K., Hastings, A., & Huxel, G. R. (1998). Weak trophic interactions and the balance of nature. *Nature, 395*, 794–798.
Nicotra, A. B., Atkin, O. K., Bonser, S. P., Davidson, A. M., Finnegan, E. J., et al. (2010). Plant phenotypic plasticity in a changing climate. *Trends in Plant Science., 15*, 684–692.
Odum, E. P. (1969). The strategy of ecosystem development. *Science, 164*, 262–270.
Pigliucci, M. (2001). *Phenotypic plasticity*. John Hopkins University Press.
Rosen, R. (1967). *Optimality principles in biology*. Butterworths.
Schoener, T. W. (1971). The theory of feeding strategies. *Annual Review of Ecology and Systematics, 2*, 369–404.
Shilov I. A. (Шилов И. А.) (1988). Biological problems of biospheric homeostasis. *Journal of General Biology (Moscow), 49*(2), 166–172. (in Russian).
Shkorbatov G. L. (Шкорбатов Г. Л.) (1971). The main adaptive features of biological systems. *Journal of General Biology (Moscow), 32*(2), 131–142. (in Russian).
Shkorbatov G. L. (Шкорбатов Г. Л.) (1982). On the construction of the general adaptation theory. *Journal of General Biology, 43*(6), 775–787. (in Russian).
Somero, G. N., Lockwood, B. L., & Tomanek, L. (2017). *Biochemical adaptation*. Sinauer Associates.
Veldhuis, M. P., Berg, M. P., Loreau, M., & Olff, H. (2018). Ecological autocatalysis: A central principle in ecosystem organization? *Ecological Monographs, 88*, 304–319.
Vermeij G. J. 2005. From phenomenology to first principles: Toward a theory of diversity. *Proceedings of the California Academy of Sciences, ser. 4*, 56 (Suppl. 1, No 2), 12–23.
Watson, A. J., & Lovelock, J. E. (1983). Biological homeostasis of the global environment: The parable of Daisyworld. *Tellus, 35B*, 286–289.
Whittaker, R. H. (1975). *Communities and ecosystems* (2nd ed.). Macmillan Publishing.
Willmer, P., Stone, G., & Johnston, I. (2005). *Environmental physiology of animals*. Blackwell Science.
Young M. W. (2002). *Circadian rhythms. Encyclopedia of life science*. Wiley. www.els.net

Chapter 12
So, Is Nature a Battlefield or a Cooperative Arena?

In Victorian England, nature was considered a perfect creation of God; it was admired up to the verge of fanaticism – that was the golden age of natural history. By replacing God with natural selection, Darwin has disappointed the romantics, albeit not completely, as he left some space for the harmony of nature and its magnificence. Modern Darwinists got rid even of this 'remnant of idealism'. No place was left for romance. I do not wish to engage in polemics with the proponents of the new 'truth', first of all, because I think that both they and their shrinking opponents are right. Sounds like a paradox, but nature indeed is both balanced and 'in flux', and I attempted to demonstrate that. The fact that nowadays a majority think otherwise makes little difference. Had a majority always been right, we would still be climbing trees today. In my opinion, also here Darwin was much closer to a correct solution. He definitely sensed the collision between the ability of organisms to upset the balance through uncontrolled reproduction and the resistance of biotic and abiotic environment to this potential threat. Intuitively, he felt where to seek an answer to the question how this collision is transformed into an ever-renewing harmony of nature.

Summarising all the above-presented material (Chaps. 3, 4, 5, 6, 7, 8, 9, 10, and 11) about the nature of relationships between organisms, it should be possible to assert that:

> With some exaggeration, one may say that all organisms both 'struggle' and cooperate simultaneously. This probably implies that a dynamic equilibrium between the forces acting to reduce biodiversity ('repulsion', 'struggle') and to increase it ('attraction', mutualism, cooperation) has been achieved.

I leave it up to my readers to decide, whether this conclusion can be interpreted as a single sufficiently meaningful 'sentence' consisting of hypotheses serving as 'words'. I can only add on my behalf that I have compiled this 'sentence' from many ideas, both my own and those of other authors. None of them could be considered undeniable truth; however, I think that, merged into one 'sentence', they must have lost most of their uncertainty.

This conclusion had crystallised in my head long ago, somewhere around 1982. I remember the sweet feeling of discovery that flushed over me back then: finally, organisms do not *either* struggle *or* cooperate, as the two opposing parties were trying to convince each other; rather, they do *both* things simultaneously. I realised that I had grasped the concept of a major collision governing the functioning of living

E. Lekevičius, *Biodiversity*, https://doi.org/10.1007/978-3-031-11582-0_12

systems and perhaps even the process of evolution. Of course, later I learned that this 'discovery' did not belong to me, that other researchers had reached similar conclusions, and some of them – many years ago. For instance, the renowned Allee (1931). Having made an overview of the entire animal kingdom, from the most primitive to the most advanced forms, he provides the following summary (pp. 355, 360–361):

> ... the potentiality of social life is inherent in living matter (...) the first step toward the development of societies had probably already been taken when life came into existence on this planet (...) the two great natural principles of struggle for existence and of co-operation are not wholly in opposition, but (...) each may have reacted upon the other in determining the trend of animal evolution.

Besides myself (see Lekevičius, 1986a, b), I think these words would attract the signatures of other authors mentioned above, such as Espinas (1878), Kropotkin (1902), Wheeler (1923), and Wilson (1975). Even Darwin himself would hardly oppose them, although he was inclined to emphasise 'struggle' as the primary driving force of evolution.

Already after 1982, I discovered that in this way I had placed myself in the camp of supporters of the balance of nature concept. I used the word 'concept', but it could as well be called an ideology, since it promotes a *belief* that nature is consistent, harmonious, sustainable, and self-regulatory. This concept was or is promoted, among others, by such conspicuous figures as S. A. Forbes, F. E. Clements, W. C. Allee, A. E. Emerson, G. E. Hutchinson, R. Lindeman, R. Margalef, E. P. Odum and H. Odum, B. C. Patten, S. J. McNaughton, L. Margulis, E.-D. Schulze, J. Lovelock, D. Pimentel, E. O. Wilson, D. Tilman, etc. Despite his concept of 'the struggle for existence', Darwin also was fascinated by nature's harmony. According to Simberloff (2014):

> Darwin nevertheless saw the ecological roles of the diversity of species as parts of an almost superorganismic nature, and his main contribution to the idea of a balance of nature was his constant emphasis on competition and other mortality factors that kept all species 'populations in check. His many metaphors and examples of the interactions among species, such as the tangled bank and the spinsters-cats-mice-bumblebees-clover stories in *The Origin of Species*, contributed to a sense of a highly balanced nature....

I could add here a well-known fact that environmental resistance also demonstrates at least a trend towards balance: averaged against time, fecundity of species usually balances out average mortality. Therefore, species can exist for millennia, without either getting extinct or becoming too abundant and supressing their competitors. Let me ask my readers: have they ever been astonished by the fact that some fish, say a cod, carelessly spreads hundreds of millions of eggs throughout its lifespan, as if aware that this is exactly the number needed for 2 ± 0.05 offspring to reach maturity and to substitute both parents (ignore the error margin; it was chosen arbitrarily and just for illustrative purpose)? On the other hand, the three-spined stickleback manages to achieve the same result with merely a few hundred eggs. Here is another example that never ceases to amaze me: a maple leave bitten by the autumn frosts is processed by decomposers for about 12 months, while a fallen pine needle is completely decomposed in a few years, but this matches the time it has spent on the pine tree. Or yet another one: as I watch a scheme presenting the biogeochemical cycle of a certain element, say sulphur, I cannot help but wonder how nature managed to construct such a sophisticated and reliable mechanism from spontaneous chemical reactions without any external help and a prior plan. If we fail to notice the touch of the almighty managers – natural selection and regulatory feedback – in these examples (and each of us could come up with many more), then we shall not be able to find it anywhere.

References

Allee, W. C. (1931). *Animal aggregations: A study in general sociology*. Chicago University Press.

Espinas, A. V. (1878). *Des Sociétés Animales*. Librairie Germer Baillière.

Kropotkin, P. (1902). *Mutual aid: A factor of evolution*. Heinemann.

Lekevičius, E. (Лекявичюс Э.). (1986a). Informational status of an ecosystem. In V. N. Maksimov (Ed.), *Ecological forecast* (pp. 157–163). Moscow University. (in Russian).

Lekevičius, E. (Лекявичюс Э.). (1986b). *Elements of a general adaptation theory*. Mokslas. (in Russsian).

Simberloff, D. (2014). The "balance of nature" – Evolution of a panchreston. *PLoS Biology, 12*(10), e1001963. https://doi.org/10.1371/journal.pbio.1001963

Wheeler, W. M. (1923). *Social life among insects*. Harcourt Brace.

Wilson, E. O. (1975). *Sociobiology* (The abridged edition). Belknap/Harvard University Press.

Part II
In Search for a Synthetic Explanation

Chapter 13
Why should an Evolutionary Biologist Start with Ecological Succession?

In the second part of the monograph, I am going to focus on the question how biodiversity occurs. Therefore, I shall seek a causal explanation. In doing that, I shall use some conclusions that I reached in Part I – I'll try to incorporate them into the explanans (explanatory part) of evolutionary facts, as promised in Chap. 2. I call such explanation synthetic, or conditionally complete (Lekevičius, 1985), because it combines functional and evolutionary regularities. I mentioned in the same chapter, that Darwin also used a similar methodology, although he himself had neither properly accentuated it nor even specified as such.

Evolutionary biologists seem to have a rather good understanding of the occurrence of species. It would suffice to mention some pretty mature concepts, such as allopatric, sympatric, parapatric, and peripatric speciation (Mayr, 1970; Grant & Grant, 2008; Futuyma, 2009), co-evolution (Ehrlich & Raven, 1964; Thompson, 1989; Futuyma, 2009), or adaptive radiation (Simpson, 1953). It is more or less clear that availability of unoccupied ecological niches promotes cladogenesis or diversification of species (Simpson, 1949, 1953). These niches can occur upon availability of new resources (1), upon immigration into a territory devoid of competitors (2), after the disappearance of predators or parasites (3), and also upon access to some key innovations that enable to eliminate competitors. Eventually, this concept has developed into the ecological theory of adaptive radiation (Schluter, 2000; Dieckmann et al., 2004). At least a few dozens of mathematical models have been produced to describe diversification as the process of filling the niches (Valentine & Walker, 1986; Chowdhury & Stauffer, 2005). Laboratory experiments with bacteria (Rainey & Travisano, 1998; Brockhurst et al., 2007) also demonstrate that diversification proceeds intensively only as long as the vacant niches, i.e. ecological opportunities, remain available. So, we already know by now that there are some centrifugal forces in nature which push species to diverge, organise themselves into guilds and thus divide tasks, disperse to the remote corners of the biosphere, and adapt to local conditions. All they need is enough of vacant space, unutilised resources, and sufficient evolvability.

And yet I daresay that we have not gone too far from the species diversification scheme presented in *The Origin of Species* (Darwin, 1872, pp. 90–91). The problem of biodiversity would have been resolved by now, if its content were that limited. There is a significant gap between the modern models of diversification, on one hand, and the picture seen by those botanists, zoologists, or ecologists who have a hands-on experience of the diversity of life forms and biotic interactions, on the

E. Lekevičius, *Biodiversity*, https://doi.org/10.1007/978-3-031-11582-0_13

other hand. Up till now, nobody has been able to answer to those who have not given up asking questions, *why* biodiversity occurred and – equally important – *how* it has become organised. It would be good to know not only how one species produces a few more but also what governs the process of diversification. As we know for sure (Chap. 3), at the macroscopic level, this governance each time results in subordination of functions, similar material cycles matching specific patterns, and standardised ecological pyramids. We have acknowledged the fact of this immense self-organisation process quite long ago, since the introduction of the concept of ecosystem. However, *how* the process looked like remains the unknown part of the history of life. This is not surprising, since, as I wrote before (Chap. 3), even the very fact of this convergence was practically forgotten, it has not been discussed for about half a century. It is certainly regrettable that we fail to find answers but even more regrettable that we might even cease posing these questions in the near future.

I aim my criticism only at the mainstream version of the theory of evolution which can be found in the latest textbooks (e.g. Maynard Smith & Szathmáry, 1995; Futuyma, 2009; Freeman & Herron, 2014). Unorthodox concepts, absent from the textbooks but not necessarily wrong, have always existed along with the 'broadly accepted' ones. Already a few decades ago, the first publications appeared which stated that macroevolution is nothing else but the process of ecosystem self-organisation by way of diversification (e.g. Stanley, 1973, 1976; Walker, 1980 – see the overview in Lekevičius, 2009). The opinion was gaining support that species diversity can be self-stimulating, because as soon as newly evolved species occupy the pre-existing niches, they themselves inevitably become vacant niches for the future consumers of these species, the herbivores, predators, or parasites. Moreover, in the course of their evolution, organisms modify their non-living environment (e.g. by depleting nutrients and discharging final products of metabolism and waste products into the environment), and then they must adapt to it themselves. The usual material cycles and ecological pyramids might have been constructed in a similar way. It seems, however, that these ideas failed to produce any significant reverberation in the mainstream version of the evolutionary theory.

Not so long ago, direct evidence was obtained to illustrate how food chains, and perhaps even ecological pyramids, are forming with the help of natural selection (Forbes et al., 2009; Hood et al., 2015). In the seventeenth century, a foreign plant was introduced into North America; it was an apple tree. Some time later, people noticed that apple trees were being attacked by an unusual pest, the local flies of the genus *Rhagoletis*. Moreover, it turned out that this was most probably a brand-new species which appeared within two or three centuries and has become specialised to feed on apples. It was also discovered that these flies (their larvae) were exploited by a parasitoid species of the genus *Diachasma*, which also was a brand-new species specialised in feeding on the new host. Different researchers have given different names to this phenomenon: phylogenetic cascade, sequential speciation, sequential divergence, autocatalytic speciation, etc.

Yet other authors (Brodersen et al., 2018), based on these and some other facts, have formulated a more general statement: 'diversification and speciation at lower trophic levels may affect radiations higher in the food web'. They tend to think, that in some cases when top predators are absent, they can evolve from smaller predators (possibly, after passing through a stage of cannibalism). To justify their position, the mentioned authors bring up, among others, the example of the adaptive radiation of haplochromine cichlids in the African Great Lakes (Victoria, Malawi, Tanganyika, and Edward). According to these authors, it is quite possible that the piscivorous cichlids of these lakes were produced by evolution from their less aggressive relatives.

These ideas might not be new (see below), but I find them really productive. On the other hand, we should not overestimate them either. Evolutionary biologists and palaeontologists have accrued much more abundant and significant information on the broadening of biodiversity spectrum, for instance, following the extinction of dinosaurs or during life's conquest of land in the Palaeozoic. We should simply use that information properly. Some methodological contemplation presented in Chap. 2, as well as the conclusions reached during the clarification of the role that biodiversity plays in the functioning of populations and communities (Part I), might help achieve this goal.

Obviously, the questions presented above are not easy. I think that one of the reasons of such a limited progress in this field of biology lies in the terminology problems. There is no agreement on the key terms necessary to any researcher pursuing similar goals. Some examples of such terms (Lekevičius, 2009) are ecological niche, adaptive zone, environmental niche, vacant/unoccupied/free niche, vacant/unoccupied/free ecospace, ökologishe Lizenz (in German), ecological opportunity, ecological release, etc. Clearly identical things are called differently, while the different ones are put under the same name. Those who prefer the term 'vacant niche', for example, tend to somewhat ignore the articles whose authors only recognise or know the terms 'free/vacant ecospace' or 'ecological opportunity' and vice versa. A deeper analysis usually reveals that all these terms refer to exactly the same phenomenon. The work of a theoretician seeking for a synthetic approach becomes somewhat of a hurdles track. As a lay physicist would put it, 'my fellow biologists, put yourselves together, start with the terms and key concepts, because there is no way to move forward without clear, monosemantic definitions'.

My interpretation of an ecological niche was presented in Chap. 4 (Fig. 4.1), and it is closer to the definition of an environmental niche, i.e. the role and place of a species within the ecosystem. In other words, in addition to the needs of a species, its ecological niche encompasses its impact and its immediate environment consisting of food resources, mutualists, competitors, enemies, and the abiotic conditions. A vacant niche hereafter should be understood as potentially suitable energetic or material resources not used by any consumer (Lekevičius, 2002, 2009). Such resources could be light, inorganic nutrients, detritus, or other living organisms, provided they are not exploited by other consumers. When vacant niches are abundant, the concept of adaptive zone is usually used (Simpson, 1949). For example, hardened lava and volcanic ash could be regarded as an adaptive zone open for colonisation by algae, cyanobacteria, lichens, and plants, but not by consumers.

Elton (1958) has once emphasised that vacant niches stimulate not only evolution but also the increase of species diversity during ecological succession. As the number of vacant niches is reduced to the minimum at the end of succession, immigrants find it harder to settle in the community. These insights have planted a logical conclusion into some heads that significant similarities exist between evolution and ecological succession (Odum, 1969; Whittaker, 1975). The idea has become even stronger after the ecosystem concept was created. The authors of this concept (Odum, 1971) discovered that global ecosystems share many common features (see Chap. 3). The greatest similarities are observed between the material turnover (cycles) and production pyramids (the 10% rule). Ecologists clearly understood that these similarities have evolved despite the fact that some local communities have been forming almost exclusively by immigration, while others – for instance, those on remote islands – have been mostly produced in situ. This implies that in both cases, succession and evolution, similar factors delimiting and shaping diversification are in play. It is not surprising, therefore, that in both cases we observe the increase of species diversity, total biomass of the ecosystems, primary production, and, most likely, efficiency of material cycles.

Unfortunately, neither ecologists nor evolutionary biologists have advanced much further from this idea. Perhaps because, traditionally, evolutionary biologists are not too interested in succession, while most ecologists show limited interest in the evolution and diversification of life (although there are exceptions, e. g. Ricklefs, 1979). I suggest breaking this established tradition and shall open the topic of diversification (cladogenesis) with a little detour towards succession. Such a strategy goes along with my chosen methodology (see Chap. 2): in metaphorical language, the laws of the evolution of life are laid right at our feet, they can be found in daily phenomena, one just hast to bring them out into the light.

References

Brockhurst, M. A., Morgan, A. D., Fenton, A., & Buckling, A. (2007). Experimental coevolution with bacteria and phage: The *Pseudomonas fluorescens*–Φ2 model system. *Infection, Genetics and Evolution, 7*, 547–552.

Brodersen, J., Post, D. M., & Seehausen, O. (2018). Upward adaptive radiation cascades: Predator diversification induced by prey diversification. *Trends in Eology and Evolution, 33*, 59–70.

Chowdhury, D., & Stauffer, D. (2005). Evolutionary ecology *in silico*: does mathematical modelling help in understanding 'generic' trends. *Journal of Biosciences, 30*, 277–287.

Darwin, C. (1872). *The Origin of Species by Means of Natural Selection* (6th ed.). John Murray (Printed in 1998 by Cox & Wyman). Reading.

Dieckmann, U., Doebeli, M., Metz, J. A. J., Tautz, D. (Eds.). (2004). *Adaptive speciation*. Cambridge University Press.

Ehrlich, P. R., & Raven, P. H. (1964). Butterflies and plants: A study of coevolution. *Evolution, 18*, 586–608.

Elton, C. (1958). *The ecology of invasions by animals and plants*. Methuen.

Forbes, A. A., Powell, T. H. Q., Stelinski, L. L., Smith, J. J., & Feder, J. L. (2009). Sequential sympatric speciation across trophic levels. *Science, 323*, 776–779.

Freeman, S., & Herron, J. C. (2014). *Evolutionary analysis* (5th ed.). Pearson.

Futuyma, D. J. (2009). *Evolution* (2nd ed.). Sinauer Associates, Inc.

Grant, P. R., & Grant, B. R. (2008). *How and why species multiply: The radiation of Darwin's finches*. Princeton University Press.
Hood, G. R., Forbes, A. A., Powell, T. H. G., Egan, S. P., & Hamerlinck, G., et al. (2015). Sequential divergence and the multiplicative origin of community diversity. *Proceedings of the National Academy of Sciences USA*. www.pnas.org/cgi/doi/10.1073/pnas.1424717112
Lekevičius, E. (Лекявичюс Э). (1985). Explanations in biology: An attempt to synthesize functional and causal approaches. In R. Pavilionis (Ed.), *Scientific knowledge: Systemic aspect* (pp. 119–135). The Academy of Sciences of the Lithuanian SSR; (in Russian).
Lekevičius, E. (2002). *The origin of ecosystems by means of natural selection*. Institute of Ecology. (Vilnius).
Lekevičius, E. (2009). Vacant niches in nature, ecology, and evolutionary theory: A mini-review. *Ekologija (Vilnius), 55*, 165–174.
Maynard, S. J., & Szathmάry, E. (1995). *The major transitions in evolution*. W. H. Freeman.
Mayr, E. (1970). *Populations, species, and evolution*. Harvard University Press.
Odum, E. P. (1969). The strategy of ecosystem development. *Science, 164*, 262–270.
Odum, E. P. (1971). *Fundamentals of ecology* (3rd ed.). W. B. Saunders.
Rainey, P. B., & Travisano, M. (1998). Adaptive radiation in a heterogeneous environment. *Nature, 394*, 69–72.
Ricklefs, R. E. (1979). *Ecology* (2nd ed.). Chiron Press.
Schluter, D. (2000). *The Ecology of adaptive radiation*. Oxford University Press.
Simpson, G. G. (1949). *The meaning of evolution*. Oxford University Press.
Simpson, G. G. (1953). *The major features of evolution*. Columbia University Press.
Stanley, S. M. (1973). An ecological theory for the sudden origin of multicellular life in the Late Precambrian. *Proceedings of the National Academy of Sciences USA, 70*, 1486–1489.
Stanley, S. M. (1976). Ideas on the timing of metazoan diversification. *Paleobiology, 2*, 209–219.
Thompson, J. N. (1989). Concepts of coevolution. *Trends in Ecology and Evolution, 4*, 179–183.
Valentine, J. W., & Walker, T. D. (1986). Diversity trends within a model taxonomic hierarchy. *Physica D: Nonlinear Phenomena, 22*, 31–42.
Walker, J. C. G. (1980). Atmospheric constraints on the evolution of metabolism. *Origins of Life., 10*, 93–104.
Whittaker, R. H. (1975). *Communities and ecosystems* (2nd ed.). Macmillan Publishing.

Chapter 14
Primary Succession as Self-Organisation of Biodiversity

Over a century ago, H. C. Cowles and V. E. Shelford described the succession of invertebrates and plants in the dunes of Lake Michigan. This study has become the classics of ecology. One would expect that primary succession must have been analysed in detail ever since. Unfortunately, even this area is still lacking clarity.

A typical primary succession also takes place in the zones of glacier retreat where the mineral bedrock is opened up. One such example is Glacier Bay in Alaska. Some 2500 km^2 of the glacier have melted within the last two centuries opening a bare layer of loam. For the last 50 years, ecologists have been studying the primary succession happening in this location (Milner et al., 2007). The studies have revealed that the ice-free area is first colonised by the pioneer species (those initiating the succession), mostly N-fixing cyanobacteria, lichens, mosses, and horsetail gametophytes. They seem to create the conditions for the arrival of other plants: not just horsetails and mosses but also Drummond's mountain avens, Sitka alders, willows, and black cottonwoods. Even later, in about a century from the beginning of succession, spruces (*Picea sitchensis*) and hemlocks (*Tsuga heterophylla*) become dominant. Thus, eventually, the area is covered by a typical forest of these latitudes, which at the beginning of succession, back in 1850, was some 20 km away from the zone opened up by the retreating glacier. There is reason to believe (Castle et al., 2016) that nutrient cycles begin to form in the early stages of such successions.

A group of Norwegian ecologists also chose the outcrops produced by retreating glaciers as their study object (Hågvar et al., 2017). They discovered that rather soon, in just 2 years, the freshly bared rock gets covered in biofilm consisting of terrestrial diatoms, accompanied by the first tiny mosses. A little later, also springtails feeding on the biofilm settle in. Water pools acquire their own life: larvae of chironomid midges foraging on the scarce detritus; some of it originates from the glacier itself, some is brought by the winds from neighbouring territories. All these processes prepare conditions for the arrival of predators: certain beetles, harvestmen, and spiders.

Naturally, the initial stages of succession involve massive loss of the immigrants, especially in the case of recently formed volcanic islands. Most of them are newcomers which cannot find their usual food or other conditions. Such destiny also awaits most of the organisms which are part of the aeolian plankton. Those include bacteria, protists, insects, arachnids, as well as spores, cysts, pollen, and seeds, all passively carried by air currents. This is a widespread phenomenon (Wilson, 1992;

E. Lekevičius, *Biodiversity*, https://doi.org/10.1007/978-3-031-11582-0_14

Thornton, 1996; Hodkinson et al., 2002). In general, the sequence of island colonisation by organisms depends exclusively on their dispersal capacity which shows no correlation with the ability to settle. Immigration is subject to chance, contrary to colonisation, which is quite predetermined. Organisms cannot survive without their niche comprising the conditions necessary for their survival. Functional dependence leads to developmental dependence.

Air flows also deliver detritus and soil particles into still uninhabited areas. Allochthonous detritus and the remains of immigrant organisms that had failed to survive could serve as the primary energy source to the arriving decomposers. The latter may be preyed upon by some predators also arriving by air, e.g. spiders and beetles. This is exactly the scenario ('heterotrophs first') defended by Hodkinson et al. (2002) on the basis of their own and other authors' empirical data. According to them, this is the beginning of soil formation in situ which prepares the conditions for plant arrival, if it is bound to happen.

One of the relatively well-explored successions is the colonisation of the small island of Krakatau or Rakata (situated between Java and Sumatra). It once hosted a tropical rainforest. However, in 1883, the island literally exploded during one of the largest volcanic eruptions in the recorded history that wiped out all of its life. Hardened lava and ash were all that remained. The island is mere a few tens of kilometres from the nearest source of species; therefore the succession was relatively fast. For instance, in 1886 the researchers already discovered 13 species of grasses and shrubs, then in 1897 there were already 49 species of plants, and by 1928 their number has approached 300. A hundred years after the catastrophe, the island sheltered over 600 invertebrate species, 30 species of land birds, including some omnivorous and predatory ones, 2 rat and 9 bat species, and the same number of reptile species (Bush & Whittaker, 1991; Thornton, 1996; Whittaker et al., 1989, 1999; Wilson, 1992).

Another quite well-examined case is the succession on Surtsey sland near the shore of Iceland. This island emerged from the ocean floor in 1963, as the result of a volcanic eruption. The first successful colonists were bacteria, fungi, mosses, and a few vascular plants (Fridriksson, 1975). Nearly half a century later, the island already sheltered 69 vascular plant species, even more mosses, as well as many protists, worms, insects, and mites typical of Iceland, plus some marine and land bird species (Frederiksen et al., 2001; Ólafsson & Ingimarsdottir, 2009; Magnusson et al., 2014). The establishment of several seagull colonies has produced an especially strong impact on the succession. The carrions of birds, the remains of their food, and their excrements facilitated the arrival of various decomposers and significantly accelerated the soil formation processes, which, in turn, promoted the increase of plant diversity. Some detritus and decomposers were brought into the island by strong winds, together with soil particles.

On neither of these islands is succession complete, so biodiversity continues to increase there. Although the island of Krakatau is covered in tropical forest, animal and plant diversity there is significantly lower than in most other forests of the region. Domination of just a few tree species (with high population densities) is, again, an unusual phenomenon for Java and Sumatra. It is anticipated that, with the arrival of new plant species, the diversity of specialised herbivores associated with them will also go up. The Krakatau community is expected to reach the stable (climax) state in a few centuries at minimum (Bush & Whittaker, 1991; Whittaker et al., 1999). For the time being, this ecosystem still has unoccupied niches, and even those already occupied presently support quite many organisms that sooner or later will have to yield them to the better adapted newcomers. A similar scenario is likely for Surtsey island as well.

The facilitation hypothesis (Connell & Slatyer, 1977; see also Tilman, 1988) states, that the first organisms, which settle on hardened lava or sand dunes, eventually change their abiotic environment in such a way that it becomes suitable for the arrival of other species. With time, the latter modify the environment even more, thus facilitating the settlement of yet new arrivals and so on. For instance, the pioneer species (cyanobacteria, algae, mosses, lichens, and grasses), together with the consumers of vegetal detritus, initiate the formation of soil, which is necessary for the establishment of other species representing later succession stages. Even though new arrivals usually displace some of the pioneer species, the overall species diversity is growing with acceleration, exponentially. This course of events (which, to the best of my knowledge, has not been denied) could be presented as the following model (Fig. 14.1).

The facilitation hypothesis explains the increase in diversity of producers; however, we should not ignore consumers either. The material presented above suggests that producers facilitate the settlement of other members of their own guild but also enable the arrival of herbivores and decomposers by themselves becoming vacant niches and by producing detritus. As the latter organisms settle in, more vacant niches are automatically created for those feeding on decomposers or herbivores. Eventually, more predators can arrive, thus closing the formation of typical production pyramids (Fig. 14.2).

For these reasons, species diversity cannot grow otherwise than with acceleration and according to the autocatalytic model. In other words, species diversity augments itself, and this process cannot be called anything else but self-organisation. I think, these conclusions could be labelled ‘community assembly rules’, even if very basic.

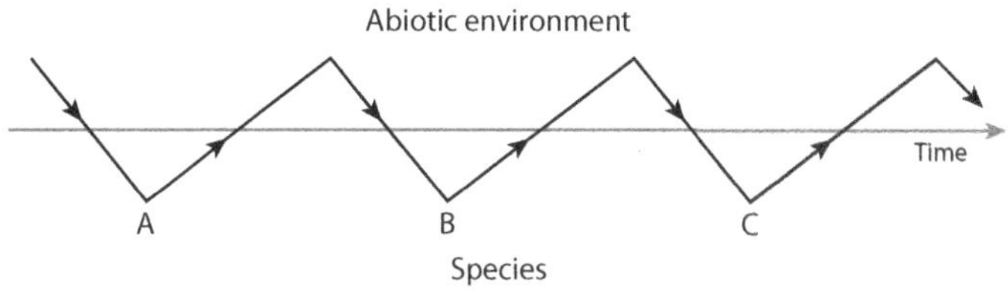

Fig. 14.1 A scheme which illustrates the facilitation hypothesis. At the onset of succession, the environmental conditions are suitable only for species A. Within a certain time, it changes the abiotic environment so that species B can also colonise the location. The latter modifies the environment even more and creates conditions for species C, etc.

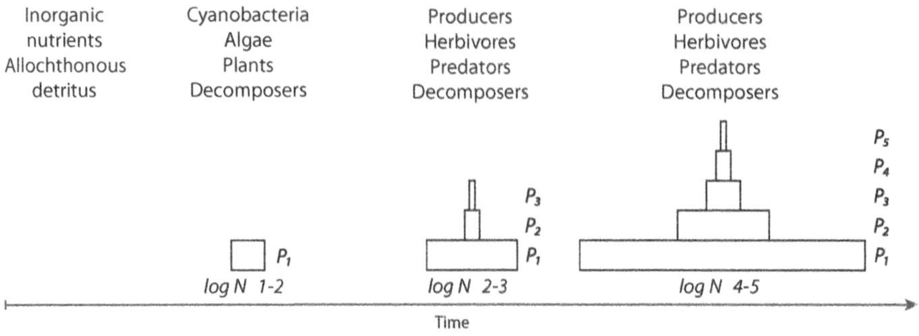

Fig. 14.2 The likely assembly stages of the consumer block during primary succession. P_i, production, grams of biomass per unit of time; *N*, species richness. Parasites are omitted for the sake of simplicity

Just as stated by Elton (1958), succession should continue as long as vacant niches are available, i.e. it stops when all suitable but unexploited resources and waste materials are depleted, and all species have their own consumers, including parasites. In theory, such situation should occur when the climax state is reached.

The classics of ecology (Odum, 1969; Whittaker, 1975) note that the primary production and total biomass of a community increase in the course of succession, along with the species diversity. I would recommend my readers to keep this conclusion in mind; it is quite important and will prove useful again and again.

References

Bush, M. B., & Whittaker, R. J. (1991). Krakatau: Colonization patterns and hierarchies. *Journal of Biogeography, 18*, 341–356.

Castle, S. C., Nemergut, D. R., Grandy, A. S., Leff, J. W., Graham, E. B., et al. (2016). Biogeochemical drivers of microbial community convergence across actively retreating glaciers. *Soil Biology and Biochemistry., 101*, 74–84.

Connell, J. H., & Slatyer, R. O. (1977). Mechanisms of succession in natural communities and their role in community stability and organization. *American Naturalist, 111*, 1119–1144.

Elton, C. (1958). *The ecology of invasions by animals and plants*. Methuen.

Frederiksen, H. B., Kraglund, H.-O., & Ekelund, F. (2001). Microfaunal primary succession on the volcanic island of Surtsey, Iceland. *Polar Research, 20*, 61–73.

Fridriksson, S. (1975). *Surtsey: Evolution of life on a Volcanic Island*. Butterworths.

Hågvar, S., Ohlson, M., & Flø, D. (2017). Animal successional pathways for about 200 years near melting glacier: A Norwegian case study. In D. Godone (Ed.), *Glacier evolution in a changing world*. InTech. https://doi.org/10.5772/intechopen.68192

Hodkinson, I. D., Webb, N. R., & Coulson, S. J. (2002). Primary community assembly on land – The missing stages: Why are the heterotrophic organisms always there first? *Journal of Ecology, 90*, 569–577.

Magnússon, B., Magnússon, S. H., Ólafsson, E., & Sigurdsson, B. D. (2014). Plant colonization, succession and ecosystem development on Surtsey with reference to neighbouring islands. *Biogeosciences, 11*, 5521–5537.

Milner, A. M., Fastie, C. L., Chapin, F. S., III, Engstrom, D. R., & Sharman, L. C. (2007). Interactions and linkages among ecosystems during landscape evolution. *Bioscience, 57*, 237–247.

Odum, E. P. (1969). The strategy of ecosystem development. *Science, 164*, 262–270.

Ólafsson, E., & Ingimarsdottir, M. (2009). The land-invertebrate fauna on Surtsey during 2002–2006. *Surtsey Research, 12*, 113–128.

Thornton, I. W. (1996). *Krakatau: The destruction and reassembly of an island ecosystem*. Harvard University Press.

Tilman, D. (1988). *Plant strategies and the dynamics and structure of plant communities*. Princeton University Press.

Whittaker, R. H. (1975). *Communities and ecosystems* (2nd ed.). Macmillan Publishing.

Whittaker, R. J., Bush, M. B., & Richards, K. (1989). Plant recolonization and vegetation succession on the Krakatau islands, Indonesia. *Ecological Monographs, 59*(2), 59–123.

Whittaker, R. J., Partomihardjo, T., & Jones, S. H. (1999). Interesting times on Krakatau: Stand dynamics in the 1990s. *Philosophical Transactions of the Royal Society of London B, 354*, 1857–1867.

Wilson, E. O. (1992). *The diversity of life*. Penguin Books.

Chapter 15
Colonisation of the Hawaiian Islands or How Evolution Complements Succession

The Hawaiian Islands is a 2650-km-long volcanic archipelago consisting of a few dozens of islands located in the tropical zone. It is some 3600 km away from the nearest continent (North America), although some small islands situated to the south and to the west of the archipelago are somewhat closer. The age of the Hawaiian Islands ranges from 5 to 0.6 million years (the Island of Hawai'i). The archipelago itself is actually over 70 million years old, but the oldest islands have submerged under the sea level again. Part of the organisms from these islands must have managed to resettle to the other parts of the archipelago and probably have survived until now. The annual precipitation is very unevenly distributed throughout the archipelago and ranges from 450 to 13,000 mm (which is one of the highest levels worldwide). The topography is also very impressive, with some summits reaching 4.2 km in height. Even on a single island, the climate may range from dry to extremely rainy at the coastal level and, on the vertical scale, from wet tropical in the low areas to temperate up in the highlands and harsh mountain climate on the highest parts (Ziegler, 2002).

Before the arrival of humans, Hawaii was completely devoid of terrestrial reptiles and amphibians, while mammals were represented by just one insectivorous bat species. As for invertebrates, usually abundant in the tropics, such groups as termites, ants, cockroaches, caddis flies, mantises, mayflies, diplurans, proturans, scorpions, pauropods, and symphylans have never managed to reach the archipelago and settle there (Howarth, 1990). All in all, representatives of 21 orders of insects common on the continents were absent from the islands (Cox et al., 1973). Most songbirds were also missing; they represented only six families (Olson, 2004). Some plant families, abundant in the tropics, are completely missing in Hawaii, while many others are represented by merely a single or a few species. On the other hand, the archipelago is extremely rich in endemic plants, including 905 species (90%) of the flowering plants alone (Price, 2009). This makes the difference between the earlier described examples of ecological succession and this particular case. The researchers have no doubts that most of the vacant niches in Hawaii were filled in by evolution before this could be done by immigration. It is considered that, with such distances to the nearest source of species, the probability of a flowering herbaceous plant arriving and settling in is about 1 per 100,000 years. Immigration probability for mammals and trees (that usually produce rather large seeds) is even lower. Of course, the time required for the arrival of microorganisms and spore plants was much shorter. The first colonists of the recently emerged islands must

E. Lekevičius, *Biodiversity*, https://doi.org/10.1007/978-3-031-11582-0_15

have been cyanobacteria, lichens, algae, mosses, and ferns (Olson, 2004). These organisms might have arrived by themselves, as spores or some other forms of the airborne plankton.

Perhaps the richest information is available on the adaptive radiation of flowering plants on the archipelago (Carlquist, 1965). One of the first arrivals some 15 million years ago were lobelias (family Campanulaceae). Presumably, they were the herbal forms. It seems (Givnish et al., 2009) a one-off immigration of just a single species was enough to produce as many as 125 endemic species representing 6 endemic genera ever since. Botanists explain such abundance of lobelias by the fact that they arrived in Hawaii earlier than their competitors like orchids, which are widespread on the continents but are represented by just a few species on the archipelago. The Hawaiian lobelias are found in almost all communities, ranging from dry shrubland to tropical rainforest. Lobelias include rosette forms growing high in the mountains, succulent forms settled on the rocky ground, but also shrubs, trees, epiphytes, and vines. This means that some herbal forms have evolved into woody plants. It should be noted that these trees have retained a number of features characteristic to herbal plans: rather soft and incompletely hardened woody tissue, short generation times (5–10 years), and limited height (up to 9 m); some species grow very fast, which is not very typical of true trees; some species bloom just once in the lifecycle and then die. So, they appear like trees but are not exactly them.

Some 5 million years ago, the first settler species from the family Asteraceae reached Hawaii, presumably from North America. Since then, it has produced 30 new species attributed by botanists to three endemic genera: *Dubautia*, *Argyroxiphium*, and *Wilkesia* (Carlquist, 1965). Just like the lobelias, their life forms are very diverse, ranging from delicate herbs, which might be similar to the hypothetical colonist, to shrubs, rosette forms, lianas, and even large trees. Even more amazing that many of these species have not lost the ability to produce cross-specific and even cross-generic hybrids (Carr & Kyhos, 1981) as demonstrated experimentally by some botanists. These 30 species, just like the local lobelias, have specialised to occupy different habitats, ranging from the arid areas with very limited precipitation, which support only herbs and meagre shrubs, to the areas with abundant rainfall covered in tropical rainforest; from the lowland species to those capable of withstanding temperate and even alpine climatic conditions (Carr & Kyhos, 1981; Baldwin & Sanderson, 1998).

Botanists have also discovered 32 endemic species of the family Caryophyllaceae representing the genus *Schiedea* on the archipelago. Again, presumably, an herbal immigrant has also produced some woody forms (Baldwin & Wagner, 2010). A similar development involved a species from the genus *Viola*, family Violaceae. Its seeds must have been transported from the Siberian tundra, presumably by migratory birds, some 3 million years ago (Ballard & Sytsma, 2000; although see Baldwin & Wagner, 2010). This fragile herb gave rise to nine new species, including some big shrubs. Botanists joke that Hawaiian forests were produced not only by lobelias but also by daisies, bellflowers, and violets (Carr & Kyhos, 1981; Baldwin & Sanderson, 1998)!

A conspicuous detail is the trend that herbal plants, upon arrival to islands, produce not only other herbal forms but also shrubs and trees. This trend has been observed not only in Hawaii but also on the island of St. Helena, the Galapagos, and many other extremely remote locations, where no tree seeds could arrive from a continent or another island (Carlquist, 1965; Ziegler, 2002). Most likely, we deal with a certain regularity here. Darwin (1872) was the first to draw attention to such a possibility. He wrote (p. 350):

> … an herbaceous plant, <…> when established on an island and having to compete with herbaceous plants alone, might readily gain an advantage by growing taller and taller and overtopping the other plants. If so, natural selection would often tend to add to the stature of herbaceous plants when growing on an island, to whatever order they belonged, and thus convert them first into bushes and ultimately into trees.

In addition to these plants, some members of genus *Metrosideros* (family Myrtaceae) somehow managed to get to Hawaii some 1.0–0.5 million years ago, and these were true trees (Price & Clague, 2002). They have also diverged after settlement, although they have managed to produce merely five endemic species adapted to different growing conditions. Nevertheless, they are abundant on the archipelago, and they even dominate the tree layer in some communities. They can grow up to 25 m tall. Botanists assume it was these genuine trees that have significantly reduced the domain of their substitute competitors hastily produced from herbal forms. Some other communities are predominated by the non-endemic acacia species, *Acacia koa* (family Fabaceae), which is also a tree whose seeds arrived in the islands relatively recently.

The formation of ecological communities on these islands has not ended with the adaptive radiation of plants. As it could be expected, they were followed by consumers. For example, Hawaii supports some 10,000 insect species produced locally, in situ, from some 350–400 colonist species (Zimmerman, 1970; Howarth, 1990; Ziegler, 2002). Perhaps the most prominent case was the radiation of fruit flies (drosophilas) (Howarth, 1990; Olson, 2004). Immigration of just one species is thought to have been sufficient to produce nearly 1000 species of endemic drosophilas within some 25 million years in Hawaii. These insects seem to have arrived in the archipelago even earlier than the blooming plants, the lobelias. One may ask what they were feeding on then. It turned out (Givnish et al., 2009) that some mosses and ferns of the genus *Diellia* might have settled on the archipelago before the drosophilas. Cyanobacteria and algae must also have been present, and, undoubtedly, the beaches were scattered with various marine detritus washed ashore by the waves. Almost all fruit flies are known to consume bacteria and fungi thriving on rotten leaves, fruits, and plant exudates. Thus, their life is inseparable from plants, although they are not herbivores in the strict sense. Moreover, it turned out that certain Hawaiian fruit fly species are associated with certain species of bacteria and fungi specific only to them, and the latter are associated with particular plants. It seems, there are no plants on the archipelago whose exudates and detritus are not exploited by some of these insects in one way or another. What is amazing is a subtle differentiation of drosophilas not just according to the types of food they consume but also according to their reaction towards the levels of light, temperature, and humidity.

Among the surviving herbivores, the most famous are Hawaiian honeycreepers (tribe Drepanidini, subfamily Carduelinae, family Fringillidae), a group of endemic birds (Ziegler, 2002). They originate from colonists close to our Eurasian bullfinch, which arrived in the islands some 7.2–5.8 million years ago. These colonists have diversified by producing 56 species, and this radiation is undoubtedly associated with the increase of lobelia diversity. This is because honeycreepers have become the main pollinators of lobelias; there are no large insects in Hawaii that usually pollinate lobelias on the continents. Besides the honeycreepers, there is another group of birds, the honeyeaters (family Meliphagidae), which plays the role of pollinators. This is a cosmopolitan family, although all the 7 species residing on the archipelago are endemic.

Speaking of the Hawaiian honeycreepers, in addition to nectar, some of their species feed exclusively on insects or fruits, while some others on seeds. To adapt to such a broad foraging niche, they had to modify the shapes of their bills to such an extent that, judging by them alone, one could expect to be dealing with representatives of several different bird orders, rather than very close relatives. Some species closely resemble African sunbirds (family Nectariniidae). However, the most peculiar of them is *Hemignathus wilsoni* (= *H. munroi*), which can still be found on the so-called Big island. This bird occupies the niche of a woodpecker. It uses its long and slender curved bill with the upper mandible twice as long as the lower one to extract larvae from beneath tree bark or self-made holes in the trunk. However, contrary to the genuine woodpeckers, it also feeds on nectar and the slender curved bill proves very useful for that. As if that were not enough, sometimes it also snatches invertebrates from the forest litter. In other words, it is a kind of 'amateur woodpecker', without a proper specialisation. It would surely loose competition with either a true woodpecker, a thrush, or a hummingbird. Actually, it reminds us of another famous species, one of Darwin's finches, which extracts larvae from cracks with the help of a cactus spine, since its beak is not fit for that purpose. According to Wilson (1992, p. 92), 'the ersatz Hawaiian woodpeckers are interesting also as an example of incomplete evolutionary convergence, since there was not enough time for this bold adaptive radiation'.

Radiation of the Hawaiian insects (not just that of drosophilas) is also surprising, because closely related species occupy niches that on the continent would usually be taken by different genera, families, or even orders (Zimmerman, 1970; Howarth, 1990; Paulay, 1994; Ziegler, 2002; Cowie & Holland, 2008). For instance, 2 orthopteran species (order Orthoptera) that somehow arrived in the islands have produced 240 new species. Among these, crickets, which feed on detritus, carrion, and fruits, are the most abundant (227 endemic species). They might have become so widespread, because neither ants nor cockroaches were present on the islands before the arrival of humans. Weevils of the genus *Proterhinus* (family Curculionidae) have also undergone radiation and now comprise, according to different sources, from 170 to 250 endemic species. They have become so abundant in the archipelago that there might not be a single tree species left whose wood and leaves are not consumed by some of these beetles. Hawaii also supports a surprisingly large number of butterflies from the endemic genus *Hyposmocoma* – about 500 or even more species. They also feed on living tissues of lichens, fungi, and plants and on vegetal detritus. A single bee species that once arrived in the archipelago has produced over 50 new species (genus *Hylaeus*), and some of them, in addition to their traditional status as pollinators, have become kleptoparasites. Hawaii also hosts locally evolved butterfly species from the widespread genus *Eupithecia* (family Geometridae). For poorly understood reasons, their divergence has been very limited, despite the presence on the islands for a few million years. Even more surprising, however, is the fact that the caterpillars of some *Eupithecia* butterflies have switched from grazing plant tissues to hunting drosophilas and other insects that accidentally touch their bristles. These exceptional predatory caterpillars seem to be globally unique. There are no other animals in Hawaii that would hunt insects in such a way, although elsewhere in tropics this type of preying is quite popular, for instance, among mantises (family Mantidae). Could this be the explanation to this anomaly? By the way, it was recently discovered that four species of the abovementioned genus *Hyposmocoma* behave in a similar but not exactly the same manner: their caterpillars weave cobwebs used to entrap snails and other molluscs abundant in Hawaii. Once the caterpillar senses that the victim is entangled, it sneaks inside the shell and devours its prey.

Wherever prey is abundant, sooner or later predators will occur. There are no ants in Hawaii and that, according to some authors, may be the reason for both the emergence of predatory caterpillars and rather high diversity of endemic ground beetles (family Carabidae), wasps, and spiders (Zimmerman, 1970; Howarth, 1990; Paulay, 1994; Ziegler, 2002). Even more surprising to zoologists are the 25 species of locally evolved dragonflies (order Odonata). The larvae of some species live in water, which is normal, but some others have evolved to survive far away from any water body, in the forest litter, where they prey on other invertebrates.

Hawaii seems to once have also supported large herbivores and predators (Ziegler, 2002; Price & Clague, 2002; Cowie & Holland, 2008; Price, 2009). Fossils and bones discovered on the islands suggest that once flightless ducks *moa-nalo* (4 species, family Anatidae) of the size of a goose were proudly marching on their ground, and wingless rails (12 species, genus *Porzana*, family Rallidae) used to shuffle the fallen leaves in search of invertebrates. An endemic buzzard used to glide high in the sky, a few owl species used to prey upon honeycreepers and other birds, and two crow species of the size of a raven would conduct their pillage raids. These predators arrived in the archipelago relatively recently; hence their diversification was modest.

What should we, as people interested in the origin of species diversity, pay a special attention to in this story? In my understanding, the colonisation of the Hawaiian Islands is an unmatched natural experiment, which tells a lot about the ways that diversity comes into being. Most probably, mosses, ferns, or lobelias, i.e. the first plants to colonise Hawaii, originally managed to settle only in those exceptional habitats that provided more or less acceptable conditions. Instead of spreading rather evenly all across the island or the entire archipelago, the colonists were able to occupy only their very limited parts because of their own specific needs and the extreme diversity of climatic and edaphic conditions. Eventually, they have spread wider, however (take a note), to achieve that they had to specialise in terms of humidity, lighting, altitude above the sea level, and, most likely, substrate and soil composition. Their evolution slowed down when the free areas and unoccupied trophic niches became less available. This may be the reason why the late arrivals from the Asteraceae and some other families have achieved much lower diversity; see also Silvertown (2004) and Tanentzap et al. (2015) for similar conclusions and additional data. It would be hardly possible to find a more compelling argument against the UNTB (Unified Neutral Theory of Biodiversity) proposed by S.P. Hubbell (see Chap. 5). In this case, species diversity was clearly determined by the diversity of environmental conditions.

The overall picture appears rather convincing. First of all, it becomes clear that the assembly stages characteristic of the Hawaiian communities have no principal differences from those observed in a typical primary succession (see Figs. 14.1 and 14.2). Colonisation starts with producers and/or detritus, followed by decomposers and herbivores; some time later, primary predators get a chance to settle on the islands as they arrive, and finally, upon attainment of a certain biodiversity level, large top predators can find enough prey to survive. Even though we have no data about when and how the soil was formed on Hawaii and the guilds of decomposers were assembled, we can assert with confidence that this happened at the early stages of community assemblage.

The stages might have been similar, but not the mechanisms – and hence not the rates – of assembling. Here, on the remote islands, the niches, similar to those of the continent or its proximate islands, are often taken by organisms from different taxonomic groups produced by evolution. Vacant niches – and they were abundant on Hawaii – promoted evolution, so it took the form of adaptive radiation. This hasty convergent evolution used the material provided by rare occasional immigrants, resulting in numerous ersatz forms: herbs have become a kind of trees; lowland plants have become alpine and vice versa; fruit flies, ground beetles, crickets, and dragonflies have occupied not only their habitual niches but also those normally taken by ants or cockroaches; some caterpillars have switched towards preying on flies and molluscs; ducks have lost the ability to fly and became a mocking version of herbivorous tortoises or ungulates; finches, in the absence of worthy competitors, have 'learned' to feed on nectar and have become pollinators, while some others have taken the vacant niche of woodpeckers. None of these endemic species is likely to stand a chance in a competition with the specialised true owners of their niches from the continent. It seems, isolation is necessary not only for the increase of the number of species but also for the preservation of their diversity. By the way, this has been proven by the catastrophic impact of introduced species on the fauna and flora of remote islands.

Hawaii is no exception to the rule; similar patterns of community assemblage are characteristic also to other islands, such as Saint Helena Island, the Galapagos, Madagascar, and New Zealand, to mention but a few (Carlquist, 1965; Wilson, 1992; Losos & Ricklefs, 2009). Their common feature is that they have been isolated from the continents for prolonged periods, which has kept the immigration rates too low to achieve biotic homogenisation. For instance, New Zealand has hosted no mammals other than bats during the latest geological periods. Birds did manage to reach it, so they have eventually evolved to occupy the niches normally taken by mammals. The radiation involved moas (11 species, large browsers), kiwis (5 species, omnivorous), parrots (including herbivorous and carrion-eating), and some enormous raptors (see the overview by Losos & Ricklefs, 2009). Madagascar separated from the African continent 88 million years ago. For many million years, it lacked many birds and mammals, which did not manage to cross the distance between the island and the continent. They were replaced by herbivorous elephant birds, ground sloths, lemurs (over 100 species, they took the niches normally occupied by monkeys), small and large carnivores related to mongooses, reptilians (over 230 endemic species), and a diverse cohort of birds (over 180 endemic species). Almost all these animals evolved in situ by way of radiation and occupied numerous highly distinctive niches (Ibid.).

My interpretation of all these facts is as follows. Species diversity has an inherent quality to attain the form of the three 'functional kingdoms' (see Chap. 3) and to produce within their framework rather invariant sets of guilds and trophic levels (the phenomenon of functional convergence of ecosystems). The inclination (or drive) of communities to achieve a predefined organisational structure is so strong that the missing chains can be hastily produced by evolution from ill-suited material, resulting in numerous peculiar ersatz species. This is what happened with the Hawaiian nature.

Of course, the topographic diversity of Hawaii has played an important role. Besides creating climatic diversity, mountains also isolate subpopulations and hinder gene flow, which leads to divergence. The same effect is presumably attained by dozens of miles of the open ocean space separating islands from one another (Grant & Grant, 2008; Futuyma, 2009). All these circumstances in Hawaii must have created more favourable conditions for divergence than those on Saint Helena or even the Galapagos, which is obvious from the results – the latter islands have not reached the same high level of endemism. The smaller distances between these islands and the mainland, compared to Hawaii, might have also contributed to that.

I would like to emphasise one regularity, which I find extremely important. The species that served as raw material for the adaptive radiation on remote islands usually do not experience any similar evolutionary perturbations in their native areas (Losos & Ricklefs, 2009). The explanation could be that, in addition to genetic variation, vacant niches are also important for divergence to occur. In a community saturated with species, a population becomes enframed and subject only to limitations, which are imposed by well-established and highly adapted species living in the nearest ecospace. In these cases, only stabilising selection is in action (Fig. 15.1). On the contrary, the same species placed into a relative ecological vacuum might have the same rates of mutation and recombination as in its original habitat; however natural selection in this case favours any, even slightest, variations that open access to some unused resources. In fact, all this is relatively well described by the

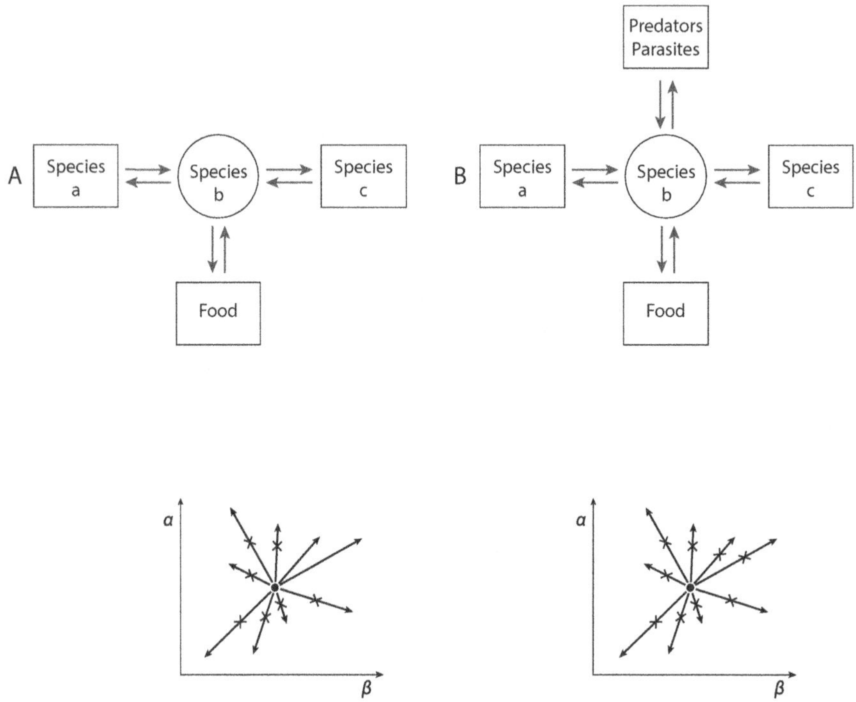

Fig. 15.1 A scheme which illustrates the statement that vacant niches represent permissions to evolve, while their absence means lack of such permissions. In situation A, species b can be seen as a vacant niche because it is unexploited. In different situation B, all the ecospace surrounding species b is occupied and devoid of vacant niches. The other two species, a and c, are competitors of species b from the same guild. The bottom part of the scheme presents the corresponding phase portraits of species b illustrating its field of evolutionary opportunities created by genetic variation. The α and β axes represent two phenotypic quantitative traits. The crossed-over arrows indicate the eliminated variations, whereas plain arrows show variations which are preserved and propagated within the population. An example of these could be cannibalistic behaviour

term *ecological release,* which implies that species are set free from their competitors and take advantage of that freedom by occupying numerous vacant niches.

In this context, some macroevolutionary mysteries may be easier to resolve. For instance, the existence of the 'living fossils'. It is considered that the existing microorganisms or other primitive organisms have changed little from the times of their appearance on the stage of evolution. Of course, one could assume that these life forms have completely lost their mutability and evolvability. However, a different explanation may be much more appropriate: they have not changed in those millions or billions of years because all their nearest (and accessible) ecospace has once been filled by their own evolutionary descendants. No matter what heritable variation would occur, none of them is able to produce a new species, which could compete with the existing ones in the local ecosystem. For example, photosynthetic sulphur bacteria could in principle mutate into cyanobacteria again; however, such a transformation would require at least a few stages, and in each of them, they would have to compete with organisms that have passed these stages long ago. The ersatz forms are viable only in the absence of competitors. Such an approach goes against the mainstream view, which postulates that the rate of evolution and divergence is determined by genetic variability.

By the way, similar arguments could be used to explain why evolution cannot repeat the indeed revolutionary breakthroughs it once has achieved. For instance, life on Earth cannot emerge from non-living matter as it did almost 4 billion years ago; prokaryotes have no second chance to produce eukaryotes nor have fishes another possibility to grow legs and settle on land. In metaphorical language, organisms were not only building bridges towards the future but they were also burning the old bridges behind them. Thus, primitive life forms have lost forever even a slightest possibility to achieve a higher level of organisation. Evolution might be able to repeat itself and produce similar forms, but only if it starts from scratch, from a complete ecological vacuum. This brings us, in my understanding, to a reasonably justified answer to the question why primitive organisms no longer evolve, so persistently sought by Darwin (see Chap. 1).

References

Baldwin, B. G., & Sanderson, M. J. (1998). Age and rate of diversification of the Hawaiian silversword alliance (Compositae). *Evolution, 95*, 9402–9406.

Baldwin, B. G., & Wagner, W. L. (2010). Hawaiian angiosperms radiations of North American origin. *Annals of Botany, 105*, 849–879.

Ballard, H. E., Jr., & Sytsma, K. J. (2000). Evolution and biogeography of the woody Hawaiian violets (*Viola*, Violaceae): Arctic origins, herbaceous ancestry and bird dispersal. *Evolution, 54*, 1521–1532.

Carlquist, S. (1965). *Island life: A natural history of the islands of the world*. The Natural History Press.

Carr, G. D., & Kyhos, D. W. (1981). Adaptive radiation in the Hawaiian silversword alliance (Compositae-Madiinae). I. Cytogenetics of spontaneous hybrids. *Evolution, 35*, 543–556.

Cowie, R. H., & Holland, B. S. (2008). Molecular biogeography and diversification of the endemic terrestrial fauna of the Hawaiian Islands. *Philosophical Transactions of the Royal Society. Series B, 363*, 3363–3376.

Cox, C. B., Healey, I. N., & Moore, P. D. (1973). *Biogeography: An ecological and evolutionary approach*. Blackwell Science Publishing.

Darwin, C. (1872). *The origin of species by means of natural selection* (6th ed.). John Murray (Printed in 1998 by Cox & Wyman).

Futuyma, D. J. (2009). *Evolution* (2nd ed.). Sinauer Associates, Inc.

Givnish, T. J., Millam, K. C., Mast, A. R., Paterson, T. B., Theim, T. J., & Hipp, A. L. (2009). Origin, adaptive radiation and diversification of the Hawaiian lobeliads (Asterales: Campanulaceae). *Proceedings of the Royal Society of London. Series B, 276*, 407–416.

Grant, P. R., & Grant, B. R. (2008). *How and why species multiply: The radiation of Darwin's finches*. Princeton University Press.

Howarth, F. G. (1990). Hawaiian terrestrial arthropods: An overview. *Bishop Museum Occasional Papers, 30*, 4–26.

Losos, J. B., & Ricklefs, R. E. (2009). Adaptation and diversification on islands. *Nature, 257*, 830–836.
Olson, S. (2004). *Evolution in Hawaii. A supplement to teaching about evolution and the nature of science*. The National Academies Press.
Paulay, G. (1994). Biodiversity on oceanic islands: Its origin and extinction. *American Zoologist, 34*, 134–144.
Price, J. (2009). Hawaiian islands, biology. In R. G. Gillespie & D. A. Claque (Eds.), *Encyclopedia of islands* (pp. 397–404). University of California Press.
Price, J. P., & Clague, D. A. (2002). How old is the Hawaiian biota? Geology and phylogeny suggest recent divergence. *Proceedings of the Royal Society of London B, 269*, 2429–2435.
Silvertown, J. (2004). The ghost of competition past in the phylogeny of island endemic plants. *Journal of Ecology, 92*, 168–173.
Tanentzap, A. J., Brandt, A. J., Smissen, R. D., Heenan, P. B., Fukami, T., & Lee, W. G. (2015). When do plant radiations influence community assembly? The importance of historical contingency in the race for niche space. *New Phytologist, 207*, 468–479.
Wilson, E. O. (1992). *The diversity of life*. Penguin Books.
Ziegler, A. C. (2002). *Hawaiian natural history, ecology, and evolution*. University of Hawai'i Press.
Zimmerman, E. C. (1970). Adaptive radiation in Hawaii with special reference to insects. *Biotropica, 2*, 32–38.

Chapter 16
Self-Assembly of Ecosystems in the Paleozoic: Overview of the Latest Sources

According to the popular press, for more than 3 billion years, all land on Earth was merely a lifeless desert battered by wind and rain. Life faced multiple obstacles to settling there. First of all, for a long period of time, all living things outside the water would be destroyed by UV radiation. Admittedly, the formation of the ozone layer a billion years or so ago had more or less eliminated that problem. Another obstacle was related to the metabolism of water and inorganic nutrients. Land organisms could not function without specialised tissues and organs dedicated to preserving water in their bodies and taking up water with dissolved nutrients from the substrate, which was weathered rocks in that case. A solution also had to be found to the problems related to the gravitation and mechanical impact of rain and wind. Finally, back then, no humus existed – it was still to be produced.

Perhaps we should start from the definition of 'land' as used by palaeontologists and evolutionary biologists in this context. Just as nowadays, back then, millions of years ago, in addition to the real 'dry land' and water bodies, there were intermediary habitats called wetlands, which could be both permanent and temporary. Palaeontological data suggest that life was resettling from water to land in gradual steps, by evolving from aquatic ancestors towards permanent wetland dwellers, then facultative wetland dwellers, till finally typical terrestrial organisms have evolved, some of them even attaining xeric (adapted to dry habitats) forms. All data about 'land-dwelling' organisms that had lived until the second half of the Carboniferous most likely refer to the life forms of these intermediary habitats (Greb et al., 2006).

In spite of that, one may still come across the assertions that life emerged in these habitats not earlier than the Silurian, some 420 million years ago, together with the first plants. It seems, however, that this opinion is quite distant from the truth. Experts, especially those applying methods of molecular phylogenetics, now are sure that life started to conquer land already in the Precambrian, some 600 million years ago or even earlier (e.g. Heckman et al., 2001; Battistuzzi & Hedges, 2009; Schaefer et al., 2010; Kenrick et al., 2012; Rota-Stabelli et al., 2013; Miyazawa et al., 2014; Wellman & Strother, 2015). On the other hand, many palaeontologists, who rely on the fossil material only, consider this dating to be unjustifiably early. For instance, fossils of millipedes (Myriapoda) are known to us only from the end of the Silurian, while molecular phylogeny annoyingly insists that this group of terrestrial organisms has come into being much earlier, in the Ordovician or even Cambrian (e.g. Miyazawa et al., 2014). Another example is molecular methods suggest that the most primitive hexapods (Hexapoda) appeared in the Early Ordovician

E. Lekevičius, *Biodiversity*, https://doi.org/10.1007/978-3-031-11582-0_16

some 479 million years ago (Misof et al., 2014), although their oldest known fossils date back only to the Devonian. Admittedly, even palaeontologists have no doubts that they must have evolved somewhat earlier, most likely in the late Silurian. Unfortunately, they have no evidence of that. However, they do hope to find it, since the undiscovered fossils have an amazing quality to get older and older – so let us wait and see. I am ashamed to admit that I have never been able to choose which side to support. I cannot get rid of the suspicion that molecular methods might prove to be correct. Hereafter, I shall mostly rely on molecular information when referring to the dates of appearance or diversification of various organisms. I should warn my readers, however, that the data provided by different authors that use molecular methods often mismatch or even contradict one another. Therefore, I myself treat the version of the course of events presented below with certain precaution and do not insist that it is 100% accurate.

One of the possible solutions that might reconcile classical palaeontologists with the supporters of modern methodologies is the assumption that life has moved onto land a very long time ago, much earlier than previously thought; however, this land was not a typical dry terrain, as we usually imagine it without any deep thinking, and not even a typical wetland, but rather yet another type of habitat, which could be called marginal or interstitial (Schaefer et al., 2010; Rota-Stabelli et al., 2013). Those could be, for instance, the shore of a sea or a lake – its surf or intertidal zone. Neither plants nor soils were necessary for the colonisation of these habitats because they were sufficiently supplied with the material washed ashore by storms and tides: small organisms, clumps of algae, carrion, etc. Whenever food is available, there will be consumers, so detritophages might have followed the detritus. It is quite possible that they simply did not fossilise because of the lack of appropriate conditions. This could be a compromise explanation based on rather untraditional reasoning.

In that case, the first terrestrial colonists might have been not plants or even lichens, as it is usually considered, but decomposers or detritophages, i.e. organisms related to modern fungi, bacteria, protists, nematodes, enchytraeids, or mites. Of course, at first, they had to adapt to somewhat different abiotic niches: the lack of water, broader fluctuation of temperatures, higher partial pressure of oxygen, increased gravitation, and more intense UV radiation. Some mycologists and palaeontologists (e.g. Gan et al., 2021) are convinced that fungi, for instance, started to colonise land at least 100 or 200 million years earlier than plants, and, prior to that, they were single-celled and exclusively aquatic organisms (Heckman et al., 2001; Porter, 2006; James et al., 2006; Beraldi-Campesi & Retallack, 2016; Berbee et al., 2017). To move onto land, fungi had to undergo significant transformation; they developed the ability to form a dense network of hyphae, started to produce hydrophobic spores that were resistant to desiccation, and could be transported by wind, while their cells began to produce pigments that protected against the UV radiation. Bacteria also had to equip themselves with similar adaptations: endospores resistant to desiccation and high temperatures or myxospores protected by thick walls. It is quite possible that by that time some fungi had already begun to form mutualistic relations with cyanobacteria and algae, although this is only speculation without any proof.

Has any of you attempted to tousle a lump (thallus) of rotten algae washed ashore while walking on the beach? Have you found a swarm of tiny white warms under it? If so, you were lucky to get a glimpse of a community very similar to that which some 500 million years ago represented a starting point in the conquest of land by life! These little warms are *Enchytraeus* species, close relatives of earthworms, which are also decomposers of phytodetritus just like them. A more thorough examination of the thallus community would most likely reveal even more life, an entire set of decomposers and predators including fungi, bacteria, and protists.

Christensen and Glenner (2010) who studied phylogenesis of the *Enchytraeus* worms (family Enchytraeidae, subclass Oligochaeta) wrote:

> Historically, the littoral zone of the sea is undoubtedly the first terrestrial or semi-terrestrial habitat where dead plant material accumulates to any greater extent, and *Enchytraeus* and *Lumbricillus* may represent early successful attempts to exploit this resource. Inland soils probably had to await the emergence of land plants in order to provide a similar food resource and here the major branch of enchytraeid genera diversified into a high number of species in the numerous decomposer networks of this varied environment.

Presently, there are over 600 enchytraeid species described, but there might be many more in the wild. Some of them remain marine species, but they are more abundant on land. These facts are also interesting because they illustrate that the organisms of the same family can adapt to living in so dramatically different environments. Apparently, the transition from the littoral zone to living in wetlands or on beaches was not too challenging for them and did not require any radical innovations.

As far as producers are concerned, their arrival and settlement on land might have been facilitated by the accumulation of detritus and humus produced by decomposers, as described above. By the way, there is no reason to think that plants had to be the first primary producers in those marginal zones or wetlands. These might have been cyanobacteria, unicellular algae, chemoautotrophs, or perhaps lichens (Wright, 1990; Beraldi-Campesi, 2013; Graham et al., 2014; Wellman & Strother, 2015). This reasoning becomes even more attractive bearing in mind that even nowadays we can find, for instance, cyanobacteria of the genus *Microcoleus* in wetlands or even dry habitats. Interestingly, these cyanobacteria can also live in water. Algae of the genus *Chlorella* and some chemosynthesisers, such as *Nitrosomonas*, also share this ability (see also Battistuzzi & Hedges, 2009). All that might be an indication of a relatively easy shift of some unicellular producers towards living in water scarcity.

By the way, palaeontologists actually find the traces of what they incline to identify as cyanobacteria and unicellular algae in the fossil soils dated over 550–1000 million years (Heckman et al., 2001; Labandeira, 2005; Battistuzzi & Hedges, 2009; Beraldi-Campesi, 2013; Wellman & Strother, 2015; Kremer et al., 2018). These soils are sometimes referred to as mats, crusts, etc. A comprehensive overview of these studies was published quite recently (Weber et al., 2016). Its authors claim that they managed to find evidence that, some 600 million years ago, biocrusts had existed in even relatively dry habitats, and they had included not only cyanobacteria and unicellular algae but, possibly, also decomposers, which suggests the likely existence of a nutrient cycle.

However, aquatic species that are closely related to terrestrial organisms can be found also in other taxons, not just among enchytraeids and unicellular producers. For instance, testate amoebae, ciliates, nematodes, rotifers, cypridids, tardigrades, and velvet worms also have aquatic relatives (Table 16.1). It is difficult to find any other interpretation of these facts, except the assumption that some organisms did not need essential changes in order to colonise land. Moreover, this colonisation must have started way before the occurrence of terrestrial plants.

Of course, the table might help us to assess the number of trophic levels that existed on land in those ancient times. But that would be a tricky task. Some of these organisms, such as ciliates and tardigrades, could be considered primary consumers, i.e. organisms feeding on algae, protists, and bacteria in this case. Of course, they could also be called phytophages, or herbivores, even if somewhat unusual ones. Among them all, only velvet warms could be considered as specialised predators foraging on multicellular prey. However, ecologists are well aware that some of the modern protozoans, rotifers, and cypridids prey upon smaller protozoans and invertebrates. By the way, not all enchytraeids and nematodes are peaceful either. So, the problem might be somewhat tricky, but we can assert, with at least some degree of certainty, that terrestrial communities of that period must have had no less than three trophic levels.

Let us move on to the Ordovician (500–435 million years ago), which is a much better explored Period. The Early Ordovician is characterised by rather intense diversification of mites (order Acari, class Arachnida) and primitive hexapods (Rota-Stabelli et al., 2013; Misof et al., 2014; Lozano-Fernandez et al., 2016, 2020). We are talking about wingless hexapods (the classification is taken from Hadley, 2018): springtails (order Collembola), proturans (order Protura), diplurans (order Diplura), jumping bristletails (order Microcoryphia), and silverfish and firebrats (order Thysanura). The last two orders of these hexapods are attributed to wingless insects (Apterygota).

Approximately at the same time, new groups of myriapods emerged: symphylans (class Symphyla) and pauropods (class Pauropoda) (although see Shear & Edgecombe, 2010). Representatives of these myriapods live also in present times, feeding mostly on detritus. Additionally, a new specialised group of predators, centipedes (class Chilopoda), are thought to have evolved among myriapods in about the same period (Rota-Stabelli et al., 2013; but see Wheat & Wahlberg, 2013).

It was in the Ordovician, some 473–471 million years ago, that liverworts, the first true plants, finally appeared in the composition of the primaeval soil (crusts and mats) (Retallack, 2009; Rubinstein et al., 2010). This assumption is supported not only by available macrofossils but also by microfossils, namely, spores. Admittedly, the recent data of molecular phylogenetics suggest some 20 million years earlier date (Morris et al., 2018). Nowadays, it has become broadly accepted that land colonisation by plants and their subsequent diversification and dispersion were facilitated by symbiotic fungi (Selosse & Strullu-Derrien, 2015; Berbee et al., 2017).

The transformation of algae into plants had taken much longer than the adaptation of unicellular algae to living in wetlands, but that should not surprise anybody. In the Late Ordovician, some 450 million years ago, lichens might have occurred in

Table 16.1 Some organisms that might have settled on the beaches and wetlands about 580–475 million years ago, i.e. before the emergence of plants (embryophytes). Also indicated here are the presumed evolutionary predecessors of these organisms ('raw material')

Group of terrestrial organisms	Their likely food sources back in those times	Probable aquatic predecessors ('raw material')
Cyanobacteria, unicellular algae, chemoautotrophs	Inorganic nutrients, light	Other representatives of the same genus or family
Myxobacteria (order Myxococcales)	Detritus	Bacteria of other orders
Protists (slime moulds and oomycetes)	Detritus, bacteria	Other slime moulds and oomycetes
Ascomycetes and imperfect fungi (divisions Ascomycota and Deuteromycota)	Detritus	Chytridiomycota (chytrids) or related groups
Testate amoebae (order Testacea, phylum Amoebozoa)	Detritus, bacteria	Naked amoebae (order Amoebina) or foraminifera (order foraminifera)
Peniculids (order Peniculida, phylum Ciliophora)	Bacteria, unicellular algae	Other representatives of this order
Terrestrial nematodes (subclass Secernentea, phylum Nematoda)	Detritus, bacteria, fungi, protozoans	Other nematodes (subclass Adenophorea)
Some rotifers of the class Eurotatoria (phylum Rotifera)	Detritus, bacteria, protozoans	Other rotifers of this class
Enchytraeids (family Enchytraeidae, subclass Oligochaeta, phylum Annelida)	Detritus, fungi, bacteria, protozoans	Other enchytraeids of this family
Terrestrial cypridids (family Cyprididae, class Ostracoda, subphylum Crustacea)	Detritus, bacteria, protozoans	Ostracods of the same or other families
Diplopods (class Diplopoda, subphylum Myriapoda)	Detritus, fungi, bacteria, protozoans	Unknown origin
Moss mites (order Oribatida, class Arachnida)	Unicellular algae, detritus, fungi, bacteria, protozoans	Other arachnids of the order Acari (?)
Tardigrades (phylum Tardigrada)	Unicellular algae, detritus, protozoans, bacteria	Other tardigrades
Velvet worms (phylum Onychophora)	Other invertebrates	Other velvet worms

The sources used include various encyclopaedic references as well as the following: Heckman et al. (2001), Porter (2006), James et al. (2006), Schaefer et al. (2010), Christensen and Glenner (2010), Kenrick et al. (2012), Rota-Stabelli et al. (2013), Wheat and Wahlberg (2013), Miyazawa et al. (2014), Misof et al. (2014), Wellman and Strother (2015), Beraldi-Campesi and Retallack (2016), and Berbee et al. (2017)

slightly drier areas (DiMichele & Hook, 1992). Of course, this is only an educated guess, since the oldest known fossils of lichens are dated some 415 million years ago (Honegger et al., 2013). If this is correct, they must have further accelerated rock erosion and formation of soil.

Since the very beginning of land colonisation, the rate of detritus decomposition must have been lower than the rate of primary production; therefore both biomass and detritus could accumulate. This created a food stock for soil-dwelling decomposers, so their diversity must have grown as well. It is quite possible (even though there is no direct evidence) that the Ordovician soils were already inhabited by typically terrestrial mutualists and decomposers produced by evolution. The modern type nutrient cycles might have also occurred back then in the wetland habitats.

The Silurian Period (444–416 million years ago) witnessed a wave of adaptive radiation of terrestrial plants, or embryophytes, albeit still a rather modest one (DiMichele & Hook, 1992; Greb et al., 2006; Wellman et al., 2013). By the end of the Silurian, mosses appeared along with liverworts, and, some 420 million years ago, they were joined by the early vascular plants Cooksonias (*Cooksonia*), which are classified as rhyniophytes. A few more million years later, the first zosterophylls, trimerophytes, and also primaeval lycopods emerged. Nevertheless, plant diversity was still very low back then, consisting of a handful of wetland-dependent species with a cosmopolitan distribution.

There are assumptions (Labandeira, 2006) that, 420–410 million years ago, some myriapods, most likely certain diplopods (class Diplopoda), have become the first terrestrial herbivores feeding on the tissues and spores of the plants of that time. C.C. Labandeira supports his assumptions both with detailed analysis of the fossils of these animals, especially their mouthparts and intestinal contents, and with the specific scratches and other markings presumably made by the herbivores that he or other researchers discovered on the vegetal fossils of that time.

Besides this group, some hexapods, which had diversified quite broadly by then (Rota-Stabelli et al., 2013; Misof et al., 2014; Lozano-Fernandez et al., 2016), such as springtails, proturans, and diplurans, as well as certain mites of that time, could also be considered herbivores, at least facultative. Indeed, there are quite many contemporary representatives of these groups that could be called herbivorous. According to C.C. Labandeira, the period of 420–410 million years ago might be considered truly revolutionary as the onset of herbivory. He thinks that these animals were mainly feeding on spores of plants (mosses and rhyniophytes) and their saps. To access the latter, they used mouth apparatus adapted to piercing soft tissues and sucking the liquid contents. Some of them were even able to nibble bits of soft stems. Some other palaeontologists go along with this opinion (see the overviews: Shear & Edgecombe, 2010; Garwood & Edgecombe, 2011; Misof et al., 2014). This suggests that, ever since, sporiferous plants ceased to be a completely vacant niche or, more precisely, an unused adaptive zone.

Let me add though, that, in my opinion, herbivory on land had occurred much earlier, if we consider herbivores also those organisms that feed on unicellular producers (see Table 16.1).

Some experts (DiMichele & Hook, 1992; Shear & Selden, 2001), however, consider that the millipedes of that time were detritophages or, in other words, decomposers. Until recently, a similar opinion was supported by almost all palaeontologists that studied this period of evolution. They all were inclined to think that herbivory had emerged only together with the first terrestrial vertebrates, or, more specifically, at the end of the Carboniferous Period when the herbivorous reptiles (synapsids) have evolved. If this version is correct, the living tissues of plants have remained unexploited for some 160 million years (as mentioned before, the first mosses appeared about 475 million years ago, while synapsids are known since 315 million years ago). Sounds unbelievable, although, allegedly, this is supported by facts.

Personally, I am inclined to share the opinion of C.C. Labandeira. I have already expressed my position earlier (Lekevičius, 2000, 2002) that many palaeontologists are wrong in thinking that herbivory was lagging behind the occurrence of terrestrial plants themselves by over 100 million years. Admittedly, my arguments were purely theoretical: such a vast vacant niche – or rather an entire set of them comprising an adaptive zone – must have attracted potential consumers as a magnet, and it could not have been so inaccessible as to remain unoccupied for so long. Even the estimated 50 million years gap between the emergence of terrestrial plants and subsequent appearance of herbivores, as suggested by C.C. Labandeira, seems too prolonged for me, and I am convinced that in the future we shall learn about much earlier origin of herbivory.

According to the fossil record, some 420–410 million years ago, new predator groups gained boost. These were trigonotarbids (order Trigonotarbida), pseudoscorpions, and the then still primitive scorpions, all from the class Arachnida. While trigonotarbids had eventually died out, pseudoscorpions and scorpions survived till modern times. Somewhat later, some 400 million years ago, harvestmen (order Opiliones) appeared on the stage – see the overviews by DiMichele and Hook (1992), Shear and Edgecombe (2010), Garwood and Edgecombe (2011), Kenrick et al. (2012), and Rota-Stabelli et al. (2013).

In the Devonian, the second wave of plant adaptive radiation rolled over the land, and it was much more intense than the previous one: it included further diversification of lycopods, zosterophylls, and trimerophytes, as well as the appearance and expansion of woody lycopods and progymnosperms (Wan et al., 2019). The last two groups might be considered the first trees originating from the herbal forms. The woody plants must have had a competitive advantage over the herbal ones since they could grow taller and overshadow their shorter rivals. A certain part of the herbal forms was probably displaced as the result, and their diversity declined. However, broadening distribution ranges of the new species resulted in their eventual diversification. Thus, the number of species produced must have been bigger than the number of species driven to extinction. Besides, a new possibility was opened for the spread and diversification of shade-tolerant forms. This led to the formation of Earth's forests with their typical layers specialised according to the light needs.

The second half of the Devonian is characterised by the expansion of horsetails (DiMichele & Hook, 1992; Wan et al., 2019), part of which has eventually developed into arborescent forms. Along with the formation of forests, the Devonian Period featured another very important turning point in the evolution of ecosystems. The occurrence of herbs, followed by woody vascular plants some time later, resulted in the increasingly dense root systems, and finally some 380–370 million years ago, the evolution of soil was essentially accomplished with the formation of turf. Ever since then, rain or wind were no longer a threat – nutrient loss to either type of erosion has been minimised.

The beginning of the Carboniferous Period (359–299 million years ago) was characterised by the appearance of the first gymnosperms, which were the 'seed ferns', or pteridosperms. These plants, apparently, could grow in relatively dry habitats. The waterlogged areas remained predominated by spore-bearing vascular plants: lycopods, horsetails, ferns, and progymnosperms. Nevertheless, the flora of that period was very poor in terms of species: only a few tree species were growing in many areas, and often just a single tree species would massively prevail over the others (Xue et al., 2015).

In the second half of the Carboniferous, most habitats were predominated by arboreal forms: woody lycopods, calamites (group of horsetails), ferns, 'seed ferns', and cordaites (primitive conifers). These forests had developed layers typical of modern forests, including lianas and probably even epiphytes (DiMichele & Hook, 1992). Plants began to move from wetlands onto a genuine dry terrain, thus filling the unoccupied niches. Let me present an illustrative fact: in the second half of the Devonian, the cosmopolitan woody progymnosperm *Archaeopteris* was highly widespread. It could be found almost everywhere, since this genus of plants was not only cosmopolitan but also predominant due to the absence of any serious rivals. However, in the second half of the Carboniferous, we already find hundreds of woody plant species, and none of the genera, with the exception of a few arborescent lycopods, could be considered dominant (Xue et al., 2015). Most probably, this illustrates how the degree of plant specialisation had been increasing along with their diversity. Mosses and spore-bearing vascular plants were prevailing in low marshlands, while the higher and drier terrain was almost exclusively occupied by gymnosperms ('seed ferns' and cordaites), a new group of plants (DiMichele et al., 2005). The occurrence of a seed must have been a powerful promotor of diversification. Since seeds are dispersed over shorter distances than spores, the degree of population isolation has increased, which had a positive effect on speciation.

Of course, occupation of the existing niches and creation of new ones were not the only process; there was enough space also for a conventional Darwinian evolution. Habitats suitable for plants must have been of limited availability, especially in the wetlands. Therefore, species were competing, 'struggling for existence', and this struggle would often result in elimination: at first, trimerophytes have displaced part of rhyniophytes and zosterophylls; later trimerophytes themselves were wiped out by herbaceous and arborescent lycopods, ferns, horsetails, and progymnosperms, until eventually, already in the Carboniferous, progymnosperms were forced to yield most of their positions to gymnosperms (Vermeij, 1987; DiMichele & Hook, 1992).

Nevertheless, more species would be produced than eliminated; therefore diversity was increasing both within local communities and at the biosphere level. It seems that animals were following the same trend. For instance, the diversity of insects, which appeared in the Carboniferous, was initially low; however, by the end of this Period, their feeding specialisation was already rather high: some of them had mouth apparatus adapted for grazing, others – for piercing, yet others – for sucking and feeding on pollen, spores, or seeds (Labandeira, 2007). The ranks of herbivores and decomposers were reinforced with orthopterans and cockroaches, as well as pulmonates. It is possible that the insects of the time were the first to test the niches of pollinators, thus facilitating the dissemination of 'seed ferns' (DiMichele & Hook, 1992).

In comparison with the Devonian, myriapods have also achieved much bigger diversity in the Carboniferous; one could even call it their 'golden age'. Some specimens of the millipedes, called arthropleurids, could grow over 2 m long. If the assumption that they were herbivorous (Schneider et al., 2010) is correct, we could imagine an amazing picture: a herd of giant millipedes grazing on herbs in a clearing of a tropical forest. If they had any enemies, these most likely were scorpions, which could also reach an impressive length of up to 75 cm (genus *Pulmonoscorpius*). Another large predator worth mentioning was a dragonfly from the genus *Meganeura*.

However, the Carboniferous is called the age of great changes for a reason. From its onset, adaptive radiation of amphibians, the first tetrapods, gained pace and eventually resulted in millions of caudate predators, representing hundreds of different species, being spread all over the forests (Carroll, 1988, 2001). While originally small, by the end of the Carboniferous, some of them have attained the length of a few metres, which suggests that the fourth and fifth trophic levels must have already existed back then. It is only natural that this happened in that particular Period when the arthropod diversity had reached such a high level. Earlier, for instance in the Ordovician or Silurian Periods, the appearance and diversification of amphibians must have been held back by the lack of food.

The most famous amphibians of that Period are labyrinthodonts. Those were mostly rather large predators reaching up to several metres in length. Originally aquatic and semi-aquatic animals, by the second half of the Carboniferous, many amphibians had become semi-terrestrial or even purely terrestrial. Along with the predatory forms, some omnivores had evolved, who were capable of consuming both animal and vegetal food, mostly seeds. Never again in their history would amphibians reach such a dominant status as in the late Carboniferous and early Permian, when they occupied most of the niches of large consumers. The extinction of gigantic arthropleurids and scorpions at the end of the Carboniferous is associated with the expansion of labyrinthodonts in exactly the same period (Carroll, 1988; DiMichele & Hook, 1992). This might have been the impact of both new aggressive predators and stronger competitors on millipedes and other invertebrates associated with them.

Some 312 million years ago, in the second half of the Carboniferous, the first still rather primitive reptiles, captorhinids or cotylosaurs, have appeared (Brocklehurst, 2017). Undoubtedly, they had some advantage over amphibians on the dry terrain from the very moment of their occurrence. Equally important was the fact that reptiles were much more agile than their amphibian competitors. Admittedly, the early reptiles seem to have been not much larger than the modern lizards, but in the subsequent Permian Period, they attained the size of the contemporary amphibians and eventually even surpassed them. Since, by then, all the niches of predators and herbivores had been occupied by amphibians, insects, or arachnids, reptiles must have had no other choice but to eliminate weaker competitors. However, back in the second half of the Carboniferous, reptiles were still playing a secondary role. For some time, their realm had been limited to the vast inland areas where the impact of amphibians was rather modest.

References

Battistuzzi, F. U., & Hedges, S. B. (2009). A major clade of prokaryotes with ancient adaptations to life on land. *Molecular and Biological Evolution, 26*, 335–343.

Beraldi-Campesi, H. (2013). Early life on land and the first terrestrial ecosystems. *Ecological Processes, 2*, 1. https://doi.org/10.1186/2192-1709-2-1

Beraldi-Campesi, H., & Retallack, G. J. (2016). Terrestrial ecosystems in the Precambrian. In B. Weber et al. (Eds.), *Biological soil crusts: An organizing principle in drylands* (Ecological studies) (Vol. 226, pp. 37–54).

Berbee, M. L., James, T. Y., & Strullu-Derrien, C. (2017). Early diverging fungi: Diversity and impact at the dawn of terrestrial life. *Annual Review of Microbiology, 71*, 41–60.

Brocklehurst, N. (2017). Rates of morphological evolution in Captorhinidae: An adaptive radiation of Permian herbivores. *PeerJ, 5*, e3200. https://doi.org/10.7717/peerj.3200

Carroll, R. L. (1988). *Vertebrate palaeontology and evolution*. W. H. Freeman and Co.

Carroll, R. L. (2001). The origin and early radiation of terrestrial vertebrates. *Journal of Paleontology, 75*, 1202–1213.

Christensen, B., & Glenner, H. (2010). Molecular phylogeny of Enchytraeidae (Oligochaeta) indicates separate invasions of the terrestrial environment. *Journal of Zoological Systematics and Evolutionary Research, 48*, 208–212.

DiMichele, W. A., & Hook, R. W. (1992). Palaeozoic terrestrial ecosystems. In A. K. Behrensmeyer et al. (Eds.), *Terrestrial ecosystems through time: Evolutionary paleoecology of terrestrial plants and animals* (pp. 205–325). University of Chicago Press.

DiMichele, W. A., Gastaldo, R. A., & Pfefferkorn, H. W. (2005). Plant biodiversity partitioning in the Late Carboniferous and Early Permian and its implications for ecosystem assembly. *Proceedings of the Californian Academy of Sciences, 56, Suppl. I*(4), 32–49.

Gan, T., Luo, T., Pang, K., et al. (2021). Cryptic terrestrial fungus-like fossils of the early Ediacaran Period. *Nature Communications, 12*, 641. https://doi.org/10.1038/s41467-021-20975-1

Garwood, R. J., & Edgecombe, G. D. (2011). Early terrestrial animals, evolution, and uncertainty. *Evolution: Education and Outreach, 4*, 489–501.

Graham, L., Lewis, L. A., Taylor, W., Wellman, C., & Cook, M. (2014). Early terrestrialization: Transition from algal to bryophyte grade. In D. T. Hanson & S. K. Rice (Eds.), *Photosynthesis in bryophytes and early land plants* (Advances in photosynthesis and respiration) (Vol. 37, pp. 9–28).

Greb, S. F., DiMichele, W. A., & Gastaldo, R. A. (2006). Evolution and importance of wetlands in earth history. In S. F. Greb & W. A. DiMichele (Eds.), *Wetlands through time* (Geological Society of America special paper) (Vol. 399, pp. 1–40). Geological Society of America.

Hadley, D. (2018). Insect classification – Subclass Apterygota. *ThoughtCo*, June 14, 2018, thoughtco.com/insect-classification-subclass-apterygota-1968651

Heckman, D. S., Geiser, D. M., Eidell, B. R., Stauffer, R. L., Kardos, N. L., & Hedges, S. B. (2001). Molecular evidence for the early colonization of land by fungi and plants. *Science, 293*, 1129–1133.

Honegger, R., Edwards, D., & Axe, L. (2013). The earliest records of internally stratified cyanobacterial and algal lichens from the Lower Devonian of the Welsh Borderland. *New Phytologist, 197*, 264–275.

James, T. Y., Kauff, F., Schoch, C. L., Matheny, P. B., Hofstetter, V., et al. (2006). Reconstructing the early evolution of fungi using a six-gene phylogeny. *Nature, 443*, 818–822.

Kenrick, P., Wellman, C. H., Schneider, H., & Edgecombe, G. D. (2012). A timeline for terrestrialization: Consequences for the carbon cycle in the Palaeozoic. *Philosophical Transactions of the Royal Society of London B, 367*, 519–536.

Kremer, B., Kaźmierczak, J., & Środoń, J. (2018). Cyanobacterial-algal crusts from Late Ediacaran paleosols of the East European Craton. *Precambrian Research, 305*, 236–246.

Labandeira, C. C. (2005). Invasion of the continents: Cyanobacterial crusts to tree-inhabiting arthropods. *Trends in Ecology and Evolution, 20*, 253–262.

Labandeira, C. C. (2006). The four phases of plant-arthropod associations in deep time. *Geologica Acta, 4*, 409–438.

Labandeira, C. (2007). The origin of herbivory on land: Initial patterns of plant tissue consumption by arthropods. *Insect Science, 14*, 259–275.

Lekevičius, E. (2000). *Only an ecosystem is alive: A somewhat nontraditional approach to the evolution of life*. Vilnius University. (in Lithuanian).

Lekevičius, E. (2002). *The origin of ecosystems by means of natural selection*. Institute of Ecology. (Vilnius).

Lozano-Fernandez, J., Carton, R., Tanner, A. R., Puttick, M. N., Blaxter, M., et al. (2016). A molecular palaeobiological exploration of arthropod terrestrialization. *Philosophical Transactions of the Royal Society of London B, 371*, 20150133.

Lozano-Fernandez, J., Tanner, A. R., Puttick, M. N., Vinther, J., Edgecombe, G. D., & Pisani, D. (2020). A Cambrian-Ordovician terrestrialization of arachnids. *Frontiers in Genetics, 11*. https://doi.org/10.3389/fgene.2020.00182

Misof, B., Liu, S., Meusemann, K., Peters, R. S., Donath, A., et al. (2014). Phylogenomics resolves the timing and pattern of insect evolution. *Science, 346*, 763–767.

Miyazawa, H., Ueda, C., Yahata, K., & Su, Z.-H. (2014). Molecular phylogeny of Myriapoda provides insights into evolutionary patterns of the mode in post-embryonoc development. *Scientific Reports, 4*, 4127.

Morris, J. L., Puttick, M. N., Clark, J. W., Edwards, D., Kenrick, P., et al. (2018). The timescale of early land plant evolution. *Proceedings of the National Academy of Sciences USA, 115*, E2274–E2283.

Porter, S. M. (2006). The Proterozoic fossil record of heterotrophic eukaryotes. In S. Xiao & A. J. Kaufman (Eds.), *Neoproterozoic geobiology and paleobiology* (pp. 1–21). Springer.

Retallack, G. J. (2009). Cambrian-Ordovician non-marine fossils from South Australia. *Alcheringa, 33*, 355–391.

Rota-Stabelli, O., Daley, A. C., & Pisani, D. (2013). Molecular timetrees reveal a Cambrian colonization of land and a new scenario for ecdysozoan evolution. *Current Biology, 23*, 392–398.

Rubinstein, C. V., Gerrienne, P., de la Puente, G. S., Astini, R. A., & Steemans, P. (2010). Early middle Ordovician evidence for land plants in Argentina (eastern Gondwana). *New Phytologist, 188*, 365–369.

Schaefer, I., Norton, R. A., Scheu, S., & Maraun, M. (2010). Arthropod colonization of land – Linking molecules and fossils in oribatid mites (Acari, Oribatida). *Molecular Phylogenetics and Evolution, 57*, 113–121.

Schneider, J. W., Lucas, S. G., Werneburg, R., & Rößler, R. (2010). Euramerican Late Pennsylvanian/Early Permian arthropleurid/tetrapod associations – Implications for the habitat and paleobiology of the largest terrestrial arthropod. In S. G. Lucas et al. (Eds.), *Carboniferous-Permian transition in Cañon del Cobre* (New Mexico Museum of Natural History and Science bulletin) (Vol. 49, pp. 49–70). New Mexico Museum of Natural History & Science.

Selosse, M. A., & Strullu-Derrien, C. (2015). Origins of the terrestrial flora: A symbiosis with fungi? *BIO Web of Conferences, 4*, 1–12.

Shear, W. A., & Edgecombe, G. D. (2010). The geological record and phylogeny of the Myriapoda. *Arthropod Structure and Development, 39*, 174–190.

Shear, W. A., & Selden, P. A. (2001). Rustling in the undergrowth: Animals in early terrestrial ecosystems. In P. G. Gensel & D. Edwards (Eds.), *Plants invade the land: Evolutionary and environmental perspectives* (pp. 29–51). Columbia University Press.

Vermeij, G. J. (1987). *Evolution and escalation: An ecological history of life*. Princeton University Press.

Wan, Z., Algeo, T. J., Gensel, P. G., Scheckler, S. E., Stein, W. E., et al. (2019). Environmental influences on the stable carbon isotopic composition of Devonian and Early Carboniferous land plants. *Palaeogeography, Palaeoclimatology, Palaeoecology, 531*, 109100. https://doi.org/10.1016/j.palaeo.2019.02.025

Weber, B., Belnap, J., & Büdel, B. (2016). Synthesis on biological soil crust research. In B. Weber et al. (Eds.), *Biological soil crusts: An organizing principle in drylands* (Vol. 2016, pp. 527–534). Springer.

Wellman, C. H., & Strother, P. K. (2015). The terrestrial biota prior to the origin of land plants (embryophytes): A review of the evidence. *Palaeontology, 58*, 603–627.

Wellman, C. H., Steemans, P., & Vecoli, M. (2013). Palaeophytogeography of Ordovician-Silurian land plants. *Geological Society of London, Memoirs, 38*, 461–476.

Wheat, C. W., & Wahlberg, N. (2013). Phylogenomic insights into the Cambrian explosion, the colonization of land and the evolution of flight in Arthropoda. *Systematic Biology, 62*, 93–109.

Wright, V. P. (1990). Terrestrialization: Soils. In D. E. G. Briggs & P. R. Crowther (Eds.), *Palaeobiology: A synthesis* (pp. 57–59). Blackwell Science.

Xue, J., Huang, P., Ruta, M., Benton, M. J., Hao, S., et al. (2015). Stepwise evolution of Paleozoic tracheophytes from South China: Contrasting leaf disparity and taxic diversity. *Earth-Science Reviews, 148*, 77–93.

Chapter 17
Self-Assembly of Ecosystems in the Paleozoic: Interpretation and Summary

In theory, terrestrial life might have come into being in the same way as it emerged in water, by way of using own resources – from inorganic compounds through chemical synthesis. This did not happen, however, presumably because the medium was by far less favourable, especially bearing in mind incessant and unpredictable variation of abiotic conditions. On the other hand, as soon as life emerged, it soon started attempts to colonise land. Not because it was short of space or resources in the aquatic environment but because the elements beyond its control would regularly eject it onto the dry terrain. Thus, from the very occurrence, organisms had been tested under unusual conditions. And we all know that the same action repeated continuously again and again eventually might lead to a positive outcome. Therefore, biologists should not be taken aback if some day they are presented with conclusive evidence that, a billion or more years ago, hundreds of unicellular producers and decomposers were spread over the shores and wetlands and created the first ecosystems there. No other characteristic of life is as conspicuous as its permanent effort to raise biomass and expand its domain.

The material presented in the last chapter (see also Lenton & Daines, 2017) might produce the impression that more or less independent local nutrient cycles might have emerged in wetlands or coastal areas very early, a billion years ago or even more. This opinion is supported by the fact that many unicellular terrestrial organisms are likely to have been 'produced' quite effortlessly from the raw material generously supplied by the waters (see Table 16.1). No wonder then that, from a chemist's perspective, these cycles are almost identical to the ones typical of the aquatic ecosystems. And, as we know, a nutrient cycle is a guarantee of a long-term stability.

Unicellular producers and decomposers had remained unexploited for some period, so they became adaptive zones open to future users. The first terrestrial consumers (or, strictly speaking, first-order biophages) most likely were peniculids closely related to the aquatic peniculids (see Table 16.1). Perhaps this could explain the apparently obvious fact that construction of ecological pyramids on land went much faster than in water (aquatic systems will be discussed a bit further). On the other hand, multicellular aquatic organisms must have been far less suitable raw material for a speedy evolution of terrestrial consumers. Quite often it was 'cheaper' and faster for life to produce a multicellular consumer from the organisms already settled on land, rather than wait for it to be created from species that were occupying similar trophic niches in aquatic communities. The aquatic precursors were useful

E. Lekevičius, *Biodiversity*, https://doi.org/10.1007/978-3-031-11582-0_17

also in this case, however, only as a raw material for the production of the ersatz samples more or less suitable for terrestrial lifestyle, whereafter evolution would take over and produce new, better-adapted forms.

Let's take some example for the sake of clarity. Evolutionary biologists that study insect phylogeny suggest the following line of their evolution (Machida, 2006; Misof et al., 2014): coneheads → springtails → two-pronged bristletails → jumping bristletails → silverfish → dragonflies. Among them, only a few species of springtails are currently attached to the aquatic environment in adult phase. One may assume (and experts indeed do) that at least the other groups that appeared later than springtails had been evolving as terrestrial organisms. However, I presented this example for another reason: in this line of evolution, only dragonflies can be considered true predators; the other listed hexapods are either decomposers or feed on producers. Dragonflies originate from silverfish, so this may be an indication that originally dragonflies might have been practicing cannibalism which eventually transformed into the predator-prey relationship. Of course, such scenario would be plausible only if, for a certain period, silverfish had been representing a vacant niche devoid of any consumers, which was a real possibility back then. By the way, we find the same regularity also in the evolutionary lineage of myriapods – the development from relatively peaceful symphylans, pauropods, and millipedes to predatory centipedes (Shear & Edgecombe, 2010; Kenrick et al., 2012; Miyazawa et al., 2014). Representatives of the last three groups seem to have no aquatic ancestors or close relatives. Thus, the in situ evolution only began with the aquatic raw material and then proceeded on land by building a new pyramid layer upon the old one. As an ecologist would add (see Chap. 6), the new trophic levels must have added extra stability to species diversity.

The proof that such scenario is at least reasonable lies in the story of the cichlids of East African lakes: Piscivorous cichlids have not arrived from other water bodies, they seem to have evolved locally from peaceful cichlids (Lekevičius, 2009; Brodersen et al., 2018). By the way, even if they are predators, they lack some features of typical piscivorous fish and are somewhat ersatz.

I hope, the Hawaiian example has already planted in our heads the realisation that evolution is not lacking in venture and even certain prankishness. Some people call it evolutionary experiments, because, allegedly, only a small part of them is successful. A mention of the peculiar Ediacaran biota and its destiny – a rapid raise and equally rapid extinction – should be enough (although this does not mean that aforesaid biota failed to produce more adapted species that had survived throughout the Cambrian, so in this case we speak not about extinction without a trace but rather about transformation into new forms of life). Ersatz species may be the very examples of such experiments. Was it possible to avoid them during the colonisation of land in the Palaeozoic? Hardly so. At least two preconditions must be met for such a species to occur: a vacant niche and a 'precursor' of its future occupant that is ill-suited for the purpose. The unicellular colonists of land that created the first terrestrial nutrient cycles might not have included ersatz species, but multicellular organisms were a different case. They must have inevitably passed through intermediate phases where they were already not aquatic life forms but not completely terrestrial either. Admittedly, neither molecular nor fossil records can offer much evidence of the existence of such species. The explanation may be that these organisms, as it should be expected of them, had existed for relatively short periods of time, so their fossils are difficult to find. They gave rather quick start to the development of new, much better-adapted species. Anyway, these might be only assumptions without any substantial proof.

As expected, the assembly of terrestrial communities has gained a momentum, since any recently formed lower trophic level would automatically become an adaptive zone for the future higher-level organisms (Table 17.1). If so, a monumental picture arises in front of us: the life of that period, especially of the Devonian and Carboniferous, was involved in the adaptive radiation wave of unprecedented scale that by far exceeded any other one known to us. How many new species adapted to wetland and terrestrial life had been produced during that period? It seems nobody has been able to answer so far; therefore, we can only guess. I, personally, think that this number reached at least few hundred thousand or even a million, i.e. a dozen or a few times less than the current number of species. Labandeira and co-authors (1997) consider that in the late Carboniferous at least the soil and forest litter contained the number of species similar to that of the present times.

Table 17.1 Vacant niches that might have promoted and directed the growth of species diversity in the Palaeozoic, as life was colonising land

BYA	Adaptive zone	Organisms settled in the adaptive zone
1.0–0.55	Sunlight, nutrients	Producers: *chemosynthesisers, cyanobacteria, algae*
	Detritus	Decomposers: *bacteria, fungi, protists*
0.55–0.47	Producers and decomposers	Primary and secondary biophages: *protists, nematodes, rotifers, tardigrades, velvet worms, mites(?)*
	Detritus	Decomposers: *bacteria, fungi, protists, nematodes, rotifers, enchytraeids, cypridids, apterygotes, mites(?)*, millipedes
0.47–0.41	Sunlight, nutrients	Producers: chemosynthesisers, cyanobacteria, algae, lichens, spore-bearing plants
	Lichens, spore-bearing plants	Herbivores: tardigrades, millipedes, mites, molluscs, insects
	Detritus	Decomposers: bacteria, fungi, protists, worms, rotifers, cypridids, apterygotes, millipedes, mites, molluscs
	Decomposers and herbivores	Predators: velvet worms, centipedes, trigonotarbids, pseudoscorpions, scorpions
0.41–0.29	Sunlight, nutrients (on *terra firma*)	Producers: spore-bearing plants and gymnosperms
	Plants	Herbivores: tardigrades, millipedes, mites, insects, molluscs, tetrapods
	Herbivores and decomposers	Secondary and tertiary predators: insects, arachnids, tetrapods

The groups listed in italic have close relatives among aquatic organisms (within the boundaries of the same genus, family, order, or class). The table was compiled almost exclusively from the reference sources indicated above (Chap. 16) and based on the concepts of vacant niche and adaptive zone

The opponents might argue by saying that such enormous biodiversity was impossible in the Carboniferous, since insect diversity alone must have been smaller by at least one order of magnitude, compared to nowadays. Plant diversity must have been much lower back then, either. What could be said in response? As far as plants and insects are concerned, the opponents would most probably be right since the burst of diversity of these organisms began in the Cretaceous and continues till now. Nevertheless, evolutionary biologists also have an explanation to the phenomenon of the diversity of plants and herbivorous insects. They say that besides the occupation of vacant niches, creation of new ones, and elimination of competitors in the course of evolution, another mechanism was in action, that of niche splitting, wherein generalists would become strictly specialised. This might be the explanation behind the development of insect (and not only their) diversity in the latest geological periods (Bush & Butlin, 2004; Janz et al., 2006; Loxdale et al., 2011; Ding et al., 2015; Nürnberg & Aberhan, 2015; Loxdale & Harvey, 2016). In addition to other mechanisms, such evolution was especially facilitated by the fact that plants had been able to produce hundreds of thousands of secondary metabolites which are thought (Loxdale et al., 2011; Becerra, 2015) to have prevented most herbivores from becoming generalists and thus promoted their diversity boom.

All the material presented above may leave one with the impression that diversity of terrestrial organisms was determined only by their feeding specialisation, in terms of nutrient cycles and energy flows. There should be no doubt, however, that variation of the climatic and edaphic conditions in time and space has also significantly contributed to the increase of species (and, most likely, genotype) diversity. One should bear in mind that, at the moment of their appearance, all species by definition are endemic, since they emerge in a small and limited territory. Later, they spread by expanding their distribution range and encounter different biotic and abiotic conditions. Thus, they not only lose their endemic status but also must specialise and adapt to the local conditions, which leads to the formation of new species. The result is greater β (regional level) and γ (continental or biome level) diversity. Once they turn in the opposite direction, which happens inevitably, they also increase α diversity (Grant & Grant, 2008). So, whenever we say that, as soon as detritus or producers appeared on land, they led towards increasing diversity of specialised decomposers or herbivores, which, in turn, initiated the occurrence of predators; this statement should be understood in the broader biogeographical context. Only such interpretation explains why we are talking about hundreds of thousands, rather than merely hundreds of new species.

Can one assume that by the end of the Carboniferous or early Permian, terrestrial communities were *completely assembled* and their functional profile has not experienced essential changes ever since? We can answer affirmatively to the last part of the question: Yes, even if there have been modifications of the functional profile of ecosystems, those changes were not substantial, as postulated by the principle of functional convergence of ecosystems (see Chap. 3). The answer to the first part, however, depends on the understanding of the word 'completely' in this context. Most ecologists would be likely to say that they associate this word with the situation wherein all potentially available resources are being used, ecological pyramids have reached a typical structure crowned by predators, and there is not a single species left free of parasites or other consumers. I would also support such position, and my only comment would be that there should be no vacant niches in assembled communities. However, this rather clear and seemingly justified definition does not imply that assembled communities are already *saturated with species*; even less so, that evolution should stop whenever the supply of vacant niches is depleted. I think that the splitting of already occupied niches should be slower than the occupation of vacant niches. Some other authors hold a similar position (e.g. Cornell, 2013; Olivares et al., 2018). I agree with these authors also on another matter: currently, we have no idea if there is any ceiling to the number of species in the biosphere, and if so, what is that number. Nonetheless, many people are convinced that such a limit should exist, although it has not been reached because of the interference by *Homo sapiens*.

Let me try to illustrate these ideas by an example. The expansion of gymnosperms in the Permian and Triassic opened a broad adaptive zone with numerous vacant trophic niches. These have eventually been occupied by herbivores, mostly insects. In the Cretaceous, prior to the appearance of the flowering plants, 95 insect families already existed, and they were feeding on the tissues, sap, and pollen of gymnosperms (Labandeira, 2014). In this period, some 125 million years ago, flowering plants came into being, and about 90 million years ago, they had already displaced from the warmer regions most ferns and gymnosperms (Berendse & Scheffer, 2009). According to Labandeira (2014), this might have been the cause of extinction of most insect species associated with spore-bearing plants and gymnosperms at the end of this Period; the monophagous species seem to have been hit the most.

On the other hand, flowering plants opened an even broader adaptive zone to herbivores. Part of insects that had been previously exploiting ferns and gymnosperms managed to switch to new hosts due to heritable variation and selection. This co-evolution lasted for a few tens of million years. Adaptive radiation of the insects associated with flowering plants was counterbalanced by the extinction of the older lineages; therefore the total diversity remained relatively unchanged throughout this entire period (125–90 million years ago, 'the Aptian-Albian gap') (Labandeira, 2014).

There is no doubt that during this period of 125 million years, diversification of herbivorous insects has been closely followed by the occurrence and evolutionary expansion of predatory insects that exploit them (Farrell, 1998). Between the Jurassic and nowadays, the number of beetle genera has increased manyfold; however, both in the Jurassic Period and now, a half or slightly more of them are herbivores, while the remaining half is more or less equally divided between the predatory and detritophagous genera (Ibid.). This might be another evidence of functional convergence. So, can we assume that in the Permian, Triassic, or Jurassic Periods (except for the episodes of mass extinction and gradual recovery) ecological communities were completely assembled in the above-mentioned sense? Rather yes than no – at least we have no reason to state the opposite. Nevertheless, they surely were not saturated with species, which is demonstrated by the steady increase of insect diversity throughout this entire period (Farrell, 1998; Ding et al., 2015) and growing degree of their specialisation, before this process was disrupted by the appearance of flowering plants.

Various experts (Wilson & Hölldobler, 2005; Janz et al., 2006; Pfennig & Pfennig, 2010; Loxdale et al., 2011; Wahlberg et al., 2013; Ding et al., 2015; Loxdale & Harvey, 2016) associate the unprecedented growth of insect diversity not only with the expansion of flowering plants and increase of their diversity but also (although see Novotny et al., 2006) with the progressively deepening specialisation of insects ('divergence of character', according to Darwin) or, in other words, with the splitting of abiotic and trophic niches. In such cases, allopatric speciation might have been supplemented by sympatric speciation, which involved disruptive selection. In both cases, genotype diversity within the boundaries of species range tended to be transformed into the diversity of ecotypes, subspecies, and, ultimately, species and families. As it was mentioned in the Preface, already Darwin had foreseen such evolutionary scenario, which postulated the crucial role of divergent, or disruptive, selection. The fact that some details of this scenario are not clear to us even today (Pfennig & Pfennig, 2010; Hautmann, 2014) is a completely different story.

Already Darwin had pointed the trend of organisms to become increasingly specialised. Simpson (1953) also has dedicated a lot of attention to this tendency. In his opinion, a perfect demonstration thereof can be mammalian evolution in the Palaeocene. Many experts consider creodonts (order Creodonta) to have been the main carnivores, while condylarths (order Condylarthra), the predecessors of ungulates, the main large herbivores of that time. However, this is not entirely correct. Both these groups belonged to the same family in early Palaeocene and shared many common features: five-toed limbs, analogous dental formulas, and even food, since both were omnivorous. Their total diversity was lower than that within most families of modern carnivores or ungulates. During the subsequent epochs, this group had split and produced different, more specialised families which had eventually evolved into different orders. Large groups evolve from smaller ones by way of divergence, concludes Simpson. Presently, we have two orders of ungulates and one of carnivores, and together they encompass 20 families. The total number of species approaches 500. Naturally, the present species are much more specialised in their feeding habits than their ancestors in the Palaeocene, although many ecological equivalents do exist.

Finally, one more rather difficult topic remains to be discussed – that of extinction. Most readers might have heard numerous statements that some 90% or even 99% of all species that have ever existed are extinct by now. Of course, no one is able to give a precise number, but even as very rough estimates, these figures are impressive. That would imply that at least hundred million species have died out! Once, I was astonished by this number: how come that just 1% of life forms have found their place on Earth, did all the extinct ones have some flaw? What was it, then?

When analysing the fossil data regarding origination and extinction rates of certain taxons, for instance, genera of carnivores or bivalves (Simpson, 1953), one gets a rather convincing picture: the rate of extinction has been just slightly slower than that of speciation throughout all geological Periods; therefore the number of genera was growing slowly. A similar turnover was characteristic to many taxons, even to those of insects, whose diversity has demonstrated exponential growth since the Carboniferous, albeit with some interruptions (Labandeira & Sepkoski, 1993; Foote, 2000; Labandeira, 2005; Nicholson et al., 2014). What might have caused such enormous loss of diversity?

One answer seems to be relatively clear – extinctions might have been caused by cosmic or geological catastrophes as well as by relatively quick and radical climate changes. As we know, the biosphere suffered the highest losses at the end of the Ordovician, Devonian, Permian, Triassic, and Cretaceous Periods. These events are often referred to as the 'big five' extinctions (although some authors count only three mass extinctions – see Bambach et al., 2004). Following the biggest catastrophes, communities would usually take 5–20 million years to be assembled again (Solé et al., 2002). Of course, the vacant adaptive zones would be filled with different organisms that were more or less equivalent to the extinct occupants of these zones. Eventually, they would usually produce more species than had existed before the extinction event. Thus, even though palaeontologists may not know some of the causes of mass extinctions, the consequences of these extinctions and the course of community recovery are understood quite well and can even be largely predicted.

Another form of extinction is pseudoextinction. This is a phenomenon, when a species undergoes transformation from generation to generation, until it eventually becomes a different species; therefore, the overall number of species remains unchanged. Intermediary forms are seldom preserved in palaeontological record, leaving the impression that the older species died out, although this is not always the case. That is exactly the process many of us imagine at the mention of the metaphor of natural selection. Presumably, many species 'died out' in this way, although their exact percentage among the extinct species is not clear. The Labandeira's data (2014) presented above and seemingly demonstrating that many species associated with gymnosperms and spore-bearing plants had gone extinct after the expansion of flowering plants could be interpreted differently: at least part of them might have gradually adapted to new hosts and transformed into different species. Of course, many old genotypes died out, but palaeontologists simply cannot register that. This could be the reason that diversity was not increasing in that period (125–90 million years ago). On the other hand, Labandeira (2005) might be right in asserting that the episodes of mass extinction of herbivorous insects in geological past should be associated with radical changes in the flora composition. The latter, undoubtedly, was influenced, among other factors, by climate change. Thus, not all species were able to switch to new hosts.

Finally, we arrive at the most controversial and, probably, the most important form of extinction that many authors refer to as background extinction. In this case, a species disappears without having left evolutionary progeny; however, its extinction is not caused by a massive natural disaster. It is widely accepted that this form of extinction has affected most of the species that have ever existed. Darwin also wrote about such extinction; he considered that its reason is usually the already mentioned struggle for existence, the interspecific competition: The stronger, better adapted species eliminate the weaker ones (Darwin, 1872, pp. 64–65, 96):

> No country can be named in which all the native inhabitants are now so perfectly adapted to each other and to the physical conditions under which they live, that none of them could anyhow be improved (…) One large group will slowly conquer another large group, reduce its numbers, and thus lessen its chance of further variation and improvement. Within the same large group, the later and more highly perfected sub-groups, from branching out and seizing on many new places in the polity of Nature, will constantly tend to supplant and destroy the earlier and less improved sub-groups. Small and broken groups and sub-groups will finally disappear.

Also (Darwin, 1872, pp. 295, 315):

> The theory of natural selection is grounded on the belief that each new variety, and ultimately each new species, is produced and maintained by having some advantage over those with which it comes into competition; and the consequent extinction of the less favoured forms almost inevitably follows (…) The inhabitants of the world at each successive period in its history have beaten their predecessors in the race for life, and are, in so far, higher in the scale, and their structure has generally become more specialised; and this may account for the common belief held by so many palaeontologists, that organisation on the whole has progressed.

It seems extinction is not random but selective; it clears the space and resources for a new, more advanced species. According to Darwin, this is the only way for evolutionary progress to be implemented, i.e. development leads from more primitive towards highly organised forms.

The main arguments of the researchers who support Darwin, including the present writer, are as follows (Lekevičius, 2018):

- The fossil record suggests that most taxons have gone extinct gradually, rather than abruptly.
- The dates of complete extinction of species, genera, and families are randomly scattered along the time axis, except for the mass extinction events.
- In most cases, a new taxon diversified not after the decline of diversity in the ecologically equivalent taxon but at approximately the same time.
- In many cases, extinction would affect the older (more primitive? less advanced?) taxons, or evolutionary predecessors, stronger than their more recent descendants.

When I presented my understanding of the colonisation of land in the Palaeozoic, I was sticking to the Darwinist position: I asserted that trimerophytes have *displaced* rhyniophytes and zosterophylls and then trimerophytes were *outcompeted* by herbaceous and arborescent lycopods, ferns, horsetails, and progymnosperms, and these have ultimately yielded their positions to gymnosperms. I was using not only the Darwinist approach but also the data collected by contemporary experts, so I feel my position was justified. However, factual evidence is scarce to support this opinion. Indeed, palaeontological data can hardly tell us about the reasons of extinctions; fossils seldom capture the nature of relationships between various organisms, especially their competition.

There is a reason to think (Carroll, 1988, 2001; DiMichele & Hook, 1992; Stanley, 1999) that reptiles had a similar impact on amphibians in the Permian and Triassic Periods. In the Permian, which followed the Carboniferous, the diversity of amphibians started to decline as reptiles were increasingly invading their habitats.

Still, during the first half of the Permian, amphibians remained at the peak of their abundance: there were 40 families, over half of them represented terrestrial forms. In the Triassic (251–200 million years ago), however, reptiles had already occupied not only most terrestrial habitats but also freshwater bodies and marine coastal zones. Aquatic habitats were inhabited by some lepidosaurs and thecodonts, as well as nothosaurs, placodonts, plesiosaurs, crocodiles, and ichthyosaurs. Naturally, the amphibians of terrestrial habitats were hit first; then the diversity of semi-terrestrial and semi-aquatic amphibians has also shrunk to the minimum. Finally, in the Late Triassic, a new wave of reptile adaptive radiation has swept over the water bodies, too, and pushed amphibians out from most aquatic habitats. All their former diversity was reduced to a meagre residue (two families). Later, already in the Cretaceous, amphibians would recover but only anurans.

Have amphibians lost competition with reptiles? Most probably, yes, but it might not be wise to stick to the doctrine of one single reason – a mistake many of us tend to commit. Abiotic changes might have also contributed to the elimination (A. V. Markov, personal communication). After all, it is well known that climate became much drier in the Permian, compared with the Carboniferous, especially in the inland areas. This might have reduced the number of habitats suitable for amphibians, which inevitably contributed to the success of reptiles. However, the impact of climate was hardly crucial or direct; it might have simply accelerated the reptilian expansion.

It seems that, by the late Carboniferous or early Permian, there were no more vacant niches left in all habitats suitable for tetrapods, except for the very dry and cold areas. Therefore, reptiles, which had started their expansion in the Permian, could not settle in these habitats in any other way but by displacing amphibians. That means that the increase of tetrapod diversity must have ceased since the Permian Period, i.e. the end of the Palaeozoic. Recently, this theoretical assumption was confirmed (Benson et al., 2016, although see Benton, 1996): it was established that rapid diversification of tetrapods in the Carboniferous and Permian Periods was replaced by a period of a relative stability in species diversity. The latter has lasted all the way to the Palaeogene, i.e. for 190 million years, despite the fact that all the old species have been replaced by new ones during that period. A diversity stagnation, presumably caused by similar reasons, can also be observed in the marine fauna, except that it had begun already in the Ordovician, some 150 million years earlier, with the arrival of large cephalopods (Sepkoski, 1979; Alroy et al., 2008).

Biologists that study background extinction have introduced the concept of 'key innovation' or evolutionary novelty, as a follow-up to some of Darwin's ideas (Lekevičius, 1986; Heard & Hauser, 1995; Schluter, 2000; Dieckmann et al., 2004; Nicholson et al., 2014). Although there is no complete agreement regarding its definition, most experts consider that key innovation can be described as heritable characteristics of organisms, certain qualities that provide their populations with exceptional possibilities to enlarge their range and, eventually, to diversify extensively. Throughout the entire evolution, organisms have acquired quite many characteristics that would match this description, at least a few hundred of them. Some examples could be vascular tissues (late Silurian trimerophytes), secondary xylem

(Middle Devonian progymnosperms), seed (early Carboniferous pteridosperms), limbs and lungs (early Carboniferous amphibians), and wings and complete metamorphosis (insects of the Carboniferous and Permian). We might be most familiar with the traits that turned our species into a supercompetitor. These are first of all the advanced brain activity and language. When describing the evolutionary consequences of key innovations, these evolutionary biologists distinguish at least two cases. One of them encompasses key innovations that enable a species to diversify earlier than the others and to occupy many vacant niches (adaptive zone). These would be most of the examples I have mentioned up till now, albeit without applying the term itself. The other case is when a guild or a trophic level is already completed, but one of its species gains the status of a supercompetitor due to some key innovation. Eventually, just as in the first case, the distribution range of that species expands, followed by an intense diversification; however, in this case, it brings along the displacement of weaker competitors. Thus, the turnover of the dominant species takes place. Of course, extinction of species is most likely to happen in the last case. The researchers think that this may be the only explanation of the disappearance of most species and higher taxons.

It might be worth also mentioning the ecologists' opinion on the matter. After all, they are the ones best positioned to know what the interspecific 'struggle for existence' is capable of and what is beyond its power. For instance, Ricklefs (1979, Chapter 35), probably the most productive author on evolution among ecologists and an acknowledged expert in both fields, has no doubts that displacement by the competitors has been one of the main reasons of species extinction in the geological past. Thirty years later, he revisited the topic (Reznick & Ricklefs, 2009) and presented additional arguments. So, concludes he and his co-author, Darwin was right also on this matter, while the same cannot be said about many representatives of the modern synthesis, who tend to deny the importance of biotic interactions in macroevolution.

To support the position of Reznik and Ricklefs (2009), I would like to present one more argument. Ever since the Permian, and possibly even the Carboniferous, land has hosted rather high number of ecologically equivalent species, sometimes referred to as ecological counterparts. As millions of years were ticking by, migration would inevitably reshuffle species on the biome or even global levels, so it would be really surprising if stronger competitors had not eliminated the weaker ones or if newcomer predators had not exterminated some of the prey species that were poorly co-adapted to live alongside.

Many evolutionary biologists of the previous century were inclined to think that the Cenozoic was characterised not only by adaptive radiation of mammals but also by a radical turnover of their fauna, caused by the pressure from more advanced competitors. The following evolutionary trends could be listed (Simpson, 1953; Carroll, 1988; Potts & Behrensmeyer, 1992): the body mass had been increasing both among the herbivores and carnivores; in the process of co-evolution, both groups had improved their agility and speed of movement; both groups were becoming increasingly specialised. There are reasons to assume that this had been accompanied by a higher degree of cephalisation, improved cognitive abilities, and enhanced social interactions. However, the rates of evolution were clearly different on various continents, and there were areas where mammalian evolution or immigration was delayed, so these trends were picked up by other vertebrate groups. Some scholars tend to think (see, e.g. Wang et al., 2008; Silvestro et al., 2015; Adams et al., 2019) that multituberculates were replaced by placental mammals, mesonychids (carnivorous ungulates) – by creodonts and carnivorous marsupials, which later were also replaced by representatives of the order Carnivora prevailing until now. It is quite possible that biotic factors alone were responsible for the extinction of large snakes (Madtsoiidae), terrestrial crocodiles (Sebecidae and Planocraniidae), and phorusrhacid ('terror') birds (Phorusrhacidae) in South America. Another example is there is almost no doubt that European passerines, which had experienced radiation in the Neogene, displaced from appropriate niches almost all non-passerine birds that had undergone radiation much earlier (Mayr, 2005).

Lately, some data has been accumulating that even the extinction of dinosaurs might have been accelerated by mammals and early birds that used to plunder poorly protected dinosaur nests with eggs or young (Bois & Mullin, 2017). This seems quite possible bearing in mind that eggs would be laid on the ground and their incubation would last for at least a few months (Erickson et al., 2017). Moreover, mammals (deltatheroideans, multituberculates, early placentals, and early marsupials) were already abundant and widespread in the Cretaceous (Luo, 2007; Wilson et al., 2012; see also Strauss, 2020), so part of them might have been serious competitors of dinosaurs and pterosaurs (Briggs, 1998). Besides the herbivores and insectivores, there were some decently sized (5–12 kg) carnivorous mammals, which were preying on other vertebrates, including small dinosaurs (Kielan-Jaworowska et al., 2004; Luo, 2007). It is quite possible that this pressure from mammals and early birds was exactly the reason why the speciation rate of dinosaurs had started to lag behind their rate of extinction some 40 million years prior to their complete disappearance (Sakamoto et al., 2016).

The researchers who support this line of thinking sometimes refer to the Great American Interchange episode. Before the Isthmus of Panama was formed, the faunas of North and South Americas had been developing in relative isolation; therefore, unique communities had emerged on each continent, with significant differences in species composition. When the isthmus occurred some 3 million years ago, the exchange of species commenced. As the result, South American mammals have suffered the most from the arriving ecological equivalents and predators (Leigh et al., 2014). The authors commented this episode as follows:

> During the Neogene, South American sloths evolved into a diversity of slow-moving herbivores, terrestrial and arboreal, with ecological analogues among Madagascar's giant lemurs which, like most of Madagascar's mammals, presumably had lower metabolism, longer lives and fewer young than ecological counterparts in more competitive continental settings such as Pleistocene North America (…) With its slow herbivores and relatively ineffective marsupial carnivores, South America's non-volant mammal fauna had become quite vulnerable to invasion (…) One would expect North American terrestrial mammals to be far more competitive than their South American counterparts, because in North America, new

> mammal species were tested by encounters with a far wider variety of competitors and predators, each of which had survived numerous similar encounters.

This example suggests that in some cases extinction of species could be caused by the immigration of unusual and highly efficient predators into the regions where they would find easily accessible, not co-adapted prey. One could also mention the decline of Australia's marsupial diversity within historic times. Beyond any doubt, humans, especially the Europeans, have significantly contributed to this extinction: a few dozens of marsupial species have most likely been drawn to extinction by feral cats, introduced foxes, rabbits and sheep, and possibly also dingoes. There is some evidence that these introduced animals either displaced their weaker local competitors, overexploited poorly adapted prey, or destroyed the habitats of resident species. Most likely, all the three factors were acting simultaneously. For a comprehensive interpretation of this unfortunate story, see Johnson (2007) and Black et al. (2012). Finally, I would not be surprised at all if some day evidence were obtained that immigration of unusual pathogens and parasites has also been responsible for a number of extinctions in the recent or distant geological past.

We could use our own example, that of humankind. The history of human origin must also have involved struggle for existence and displacements (Foley, 1987). There is sufficient evidence to claim (Haile-Selassie et al., 2016) that some 3.8–3.0 million years ago, at least four species of hominins coexisted in East Africa. They occupied similar niches, and some even had overlapping ranges in certain periods of time. The same could be said about the former coexistence of *Homo erectus* and *Australopithecus*, our species and Neanderthals. Most likely, the relationships between these hominins had not always been peaceful. Tens of thousands of years ago, only one, undoubtedly, the most competitive species, was left of all these genera and species (Flinn et al., 2005). So, war may be an invention of hominins; it is ancient and does not belong to our species. This especially aggressive group of primates failed to eliminate only apes (superfamily Hominoidea) that resided in forests and poor or inaccessible locations; thus, their diversity remains relatively high (25 species).

However, already in the second half of the last century, a new conjuncture started to develop in evolutionary biology and has become prevailing. With the establishment of neo-Darwinism, everybody became focused on differential *reproduction of individuals (genotypes, genes)*, while the question of differential *extinction of species* was abandoned, well, not completely – since palaeontologists seem to have taken it over from biologists. And their position was quite different from that of Darwin. This approach might be attributed to G.G. Simpson, one of the most outstanding palaeontologists. He summed up his position as follows:

> … the usual sequence is for one dominant group to die out, leaving the zone empty, before the other group becomes abundant …. (Simpson, 1944, p. 212)

So, this researcher was of the opinion that extinction of species could be caused by other factors than the biotic ones. Such attitude remains popular also among the modern generation of palaeontologists (Marshall et al., 1982; Benton, 1987, 1996; Rampino et al., 1997; Valkenburgh, 1999; Eldredge, 2003; Wesley-Hunt, 2005;

Sahney et al., 2010; Prevosti et al., 2013; MacLeod, 2014). In their understanding, the role of competition – and thus of natural selection – in macroevolution has been rather unimportant; biotic factors might produce a significant impact only on small temporal and spatial scales, i.e. microevolution, while abiotic factors determine long-term and global changes in life. They think that the geological past contains not only large-scale catastrophes that resulted in mass extinctions but also a series of smaller ones, such as perpetual climate changes, volcanic eruptions, continental drift, fluctuations of solar activity, and impacts of relatively small asteroids. These mini catastrophes would eliminate certain taxons and clear niches for other taxons that had been overshaded previously. Consequently, 'passive replacement' must have been much more common throughout the course of life's evolution than competitive displacement (= competitive exclusion). One may even encounter statements that life would not be able to evolve without abiotic disasters.

In my opinion, such disregard of biotic interactions is an obvious mistake. Even if we accept that the primary causes of extinction of species and larger taxons in many cases indeed were disasters of abiotic nature, we still must acknowledge another fact: as certain species died out, they would drag along a vast set of other species linked with the extinct ones through trophic or other relations (co-extinctions, e. g. Vermeij, 2004; Galetti et al., 2017). Species are independent of each other neither in life nor in death.

Darwin would find it hard to believe that we, his descendants, have gone so far away (in a positive or negative sense?) from the idea of natural selection. And this attitude of palaeontologists seems to gain support among certain biologists of great authority who set the fashion in evolutionary biology (e.g. Futuyma, 2009). Admittedly, merely a few decades ago, some other evolutionary biologists were still sticking to the position close to that of Darwin when discussing the reasons of species extinction (e.g. Mayr, 1963; Bock, 1972).

The polemics on extinction continues, and its outcome is far from clear (Voje et al., 2015). Recently, there have been some indications that both seemingly contradicting positions regarding the causes of background extinctions, which were concisely presented above, might turn out to be correct: the local and global physical changes might have indirectly affected the outcomes of interspecific competition (Fortelius et al., 2014). Whatever is the end of this story, I am convinced that it will determine the future of Darwinism and evolutionary biology.

There are two possible extreme resolutions. The first one is there was no selectivity in extinction, evolution has been driven by random events, and therefore Darwinism, as the theory of macroevolution, is mostly obsolete. The second one is selectivity was involved, and it was ensured by the Darwinian struggle for existence.

References

Adams, N. F., Rayfield, E. J., Cox, P. G., Cobb, S. N., & Corfe, I. J. (2019). Functional tests of the competitive exclusion hypothesis for multituberculate extinction. *Royal Society Open Science, 6*, 181536.

Alroy, J., Aberhan, M., Bottjer, D. J., Foote, M., Fürsich, F. T., et al. (2008). Phanerozoic trends in the global diversity of marine invertebrates. *Science, 321*, 97–100.

Bambach, R. K., Knoll, A. H., & Wang, S. C. (2004). Origination, extinction, and mass depletions of marine diversity. *Paleobiology, 30*, 522–542.

Becerra, J. X. (2015). On the factors that promote the diversity of herbivorous insects and plants in tropical forests. *Proceedings of the National Academy of Sciences USA, 112*, 6098–6103.

Benson, R. B. J., Butler, R. J., Alroy, J., Mannion, P. D., Carrano, M. T., & Lloyd, G. T. (2016). Near-stasis in the long-term diversification of Mesozoic tetrapods. *PLoS Biology, 14*(1), e1002359. https://doi.org/10.1371/journal.pbio.1002359

Benton, M. J. (1987). Progress and competition in macroevolution. *Biological Reviews, 62*, 305–338.

Benton, M. J. (1996). On the nonprevalence of competitive replacement in the evolution of tetrapods. In D. Jablonski et al. (Eds.), *Evolutionary paleobiology* (pp. 185–210). University of Chicago Press.

Berendse, F., & Scheffer, M. (2009). The angiosperm radiation revisited, an ecological explanation for Darwin's 'abominable mystery'. *Ecology Letters, 12*, 865–872.

Black, K. H., Archer, M., Hand, S. J., & Godthelp, H. (2012). The rise of Australian marsupials: A synopsis of biostratigraphic, phylogenetic, palaeoecologic and palaeobiogeographic understanding. In J. A. Talent (Ed.), *Earth and life: Global biodiversity, extinction intervals and biogeographic perturbations trough time* (pp. 983–1078). Springer.

Bock, W. J. (1972). Species interactions and macroevolution. *Evolutionary Biology, 5*, 1–24.

Bois, J., & Mullin, S. J. (2017). Dinosaur nest ecology and predation during the Late Cretaceous: Was there a relationship between upper Cretaceous extinction and nesting behavior? *Historical Biology, 29*, 976–986.

Briggs, J. C. (1998). Biotic replacements – Extinction or clade interaction? *Bioscience, 48*, 389–395.

Brodersen, J., Post, D. M., & Seehausen, O. (2018). Upward adaptive radiation cascades: Predator diversification induced by prey diversification. *Trends in Ecology and Evolution, 33*, 59–70.

Bush, G. L., & Butlin, R. K. (2004). Sympatric speciation in insects. In U. Dieckmann, M. Doebeli, J. A. J. Metz, & D. Tautz (Eds.), *Adaptive speciation* (pp. 229–248). Cambridge University Press.

Carroll, R. L. (1988). *Vertebrate palaeontology and evolution*. W. H. Freeman and Co.

Carroll, R. L. (2001). The origin and early radiation of terrestrial vertebrates. *Journal of Paleontology, 75*, 1202–1213.

Cornell, H. V. (2013). Is regional species diversity bounded or unbounded? *Biological Reviews, 88*, 140–165.

Darwin, C. (1872). *The origin of species by means of natural selection* (6th ed.). John Murray (Printed in 1998 by Cox & Wyman).

Dieckmann, U., Doebeli, M., Metz, J. A. J., & Tautz, D. (Eds.). (2004). *Adaptive speciation*. Cambridge University Press.

DiMichele, W. A., & Hook, R. W. (1992). Paleozoic terrestrial ecosystems. In A. K. Behrensmeyer et al. (Eds.), *Terrestrial ecosystems through time: Evolutionary paleoecology of terrestrial plants and animals* (pp. 205–325). University of Chicago Press.

Ding, Q., Labandeira, C. C., Meng, Q., & Ren, D. (2015). Insect herbivory, plant-host specialization and tissue partitioning on mid-Mesozoic broadleaved conifers of Northeastern China. *Palaeogeography, Palaeoclimatology and Palaeoecology, 440*, 259–273.

Eldredge, N. (2003). The sloshing bucket: How the physical realm controls evolution. In J. Crutchfield & P. Schuster (Eds.), *Evolutionary dynamics: Exploring the interplay of selection, accident, neutrality, and function* (pp. 3–30). Oxford University Press.
Erickson, G. M., Zelenitsky, D. K., Kay, D. I., & Norell, M. A. (2017). Dinosaur incubation periods directly determined from growth-line counts in embryonic teeth show reptilian-grade development. *Proceedings of the National Academy of Sciences USA, 114*, 540–545.
Farrell, B. D. (1998). "Inordinate fondness" explained: Why are there so many beetles? *Science, 281*, 555–559.
Flinn, M. V., Geary, D. C., & Ward, C. V. (2005). Ecological dominance, social competition, and coalitionary arms races: Why humans evolved extraordinary intelligence. *Evolution and Human Behavior, 26*, 10–46.
Foley, R. (1987). *Another unique species: Patterns in human evolutionary ecology*. Longman Group.
Foote, M. (2000). Origination and extinction components of diversity: General problems. In D. H. Erwin & S. L. Wing (Eds.), *Deep time: Paleobiology's perspective* (pp. 74–102). The Paleontological Society, cop.
Fortelius, M., Eronen, J. T., Kaya, F., Tang, H., Raia, P., & Puolamäki, K. (2014). Evolution of Neogene mammals in Eurasia: Environmental forcing and biotic interactions. *Annual Review of Earth and Planetary Sciences, 42*, 579–604.
Futuyma, D. J. (2009). *Evolution* (2nd ed.). Sinauer Associates.
Galetti, M., Moleon, M., Jordano, P., Pires, M. M., Guimaraes, P. R., Jr., et al. (2017). Ecological and evolutionary legacy of megafauna extinctions. *Biological Reviews, 93*, 845–862.
Grant, P. R., & Grant, B. R. (2008). *How and why species multiply: The radiation of Darwin's finches*. Princeton University Press.
Haile-Selassie, Y., Melillo, S. M., & Su, D. F. (2016). The Pliocene hominin diversity conundrum: Do more fossils mean less clarity. *Proceedings of the National Academy of Sciences USA, 113*, 6364–6371.
Hautmann, M. (2014). Diversification and diversity partitioning. *Paleobiology, 40*, 162–176.
Heard, S. B., & Hauser, D. L. (1995). Key evolutionary innovations and their ecological mechanisms. *Historical Biology, 10*, 151–173.
Janz, N., Nylin, S., & Wahlberg, N. (2006). Diversity begets diversity: Host expansions and the diversification of plant-feeding insects. *BMC Evolutionary Biology, 6*, 4. https://doi.org/10.1186/1471-2148-6-4
Johnson, C. (2007). *Australia's mammal extinctions: A 50000 year history*. Cambridge University Press.
Kenrick, P., Wellman, C. H., Schneider, H., & Edgecombe, G. D. (2012). A timeline for terrestrialization: Consequences for the carbon cycle in the Palaeozoic. *Philosophical Transactions of the Royal Society of London B, 367*, 519–536.
Kielan-Jaworowska, Z., Cifelli, R. L., & Luo, Z.-X. (2004). *Mammals from the age of dinosaurs: Origins, evolution, and structure*. Columbia University Press.
Labandeira, C. C. (2005). The fossil record of insect extinction: New approaches and future directions. *American Entomologist, 51*, 14–29.
Labandeira, C. (2014). Why did terrestrial insect diversity not increase during the angiosperm radiation? Mid-Mesozoic, plant-associated insect lineages harbor clues. In P. Pontarotti (Ed.), *Evolutionary biology: Genome evolution, speciation, coevolution and origin of life*. Springer.
Labandeira, C. C., & Sepkoski, J. J., Jr. (1993). Insect diversity in the fossil record. *Science, 261*, 310–315.
Labandeira, C. C., Phillips, T. L., & Norton, R. A. (1997). Oribatid mites and the decomposition of plant tissues in Paleozoic coal-swamp forests. *PALAIOS, 12*, 318–353.
Leigh, E. G., O'Dea, A., & Vermeij, G. J. (2014). Historical biogeography of the isthmus of Panama. *Biological Reviews, 89*, 148–172.
Lekevičius, E. (Лекявичюс Э). (1986). *Elements of a general adaptation theory*. Mokslas. (in Russsian).

Lekevičius, E. (Лекявичюс Э). (2009). On some analogies between ecosystem evolution and development of economy: From A. Smith and C. Darwin to the latest ideas. In L. E. Grinin, A. V. Markov, & A. V. Korotaev (Eds.), *Evolution: Cosmic, biological, and social* (pp. 226–260). Librokom. (in Russian).
Lekevičius, E. (Лекявичюс Э). (2018). Biodiversity – What for and why? II. In search for a synthetic explanation. *Journal of General Biology (Moscow), 79*, 221–236. (in Russian).
Lenton, T. M., & Daines, S. J. (2017). Matworld – The biogeochemical effects of early life on land. *New Phytologist, 215*, 531–537.
Loxdale, H. D., & Harvey, J. A. (2016). The 'generalism' debate: Misinterpreting the term in the empirical literature focusing on dietary breadth in insects. *Biological Journal of the Linnean Society, 119*, 265–282.
Loxdale, H. D., Lushai, G., & Harvey, J. A. (2011). The evolutionary improbability of 'generalism' in nature, with special reference to insects. *Biological Journal of the Linnean Society, 103*, 1–18.
Luo, Z.-X. (2007). Transformation and diversification in early mammal evolution. *Nature, 450*, 1011–1019.
Machida, R. (2006). Evidence from embryology for reconstructing the relationships of hexapod basal clades. *Arthropod Systematics and Phylogeny, 64*, 95–104.
MacLeod, N. (2014). The geological extinction record: History, data, biases, and testing. In G. Keller & A. C. Kerr (Eds.), *Volcanism, impacts, and mass extinctions: Causes and effects* (Geological Society of America. Special paper) (Vol. 505, pp. 1–28). The Geological Society of America.
Marshall, L. G., Webb, S. D., Sepkoski, J. J., & Raup, D. M. (1982). Mammalian evolution and the Great American Interchange. *Science, 215*, 1351–1357.
Mayr, E. (1963). *Animal species and evolution*. Harvard University Press.
Mayr, G. (2005). Paleogene fossil record of birds in Europe. *Biological Reviews, 80*, 515–542.
Misof, B., Liu, S., Meusemann, K., Peters, R. S., Donath, A., et al. (2014). Phylogenomics resolves the timing and pattern of insect evolution. *Science, 346*, 763–767.
Miyazawa, H., Ueda, C., Yahata, K., & Su, Z.-H. (2014). Molecular phylogeny of Myriapoda provides insights into evolutionary patterns of the mode in post-embryonic development. *Scientific Reports, 4*, 4127.
Nicholson, D. B., Ross, A. J., & Mayhew, P. J. (2014). Fossil evidence for key innovations in the evolution of insect diversity. *Proceedings of the Royal Society B, 281*, 20141823.
Novotny, V., Drozd, P., Miller, S. E., Kulfan, M., Janda, M., et al. (2006). Why are there so many species of herbivorous insects in tropical rainforests? *Science, 313*, 1115–1118.
Nürnberg, S., & Aberhan, M. (2015). Interdependence of specialization and biodiversity in Phanerozoic marine invertebrates. *Nature Communications, 6*, 1–8.
Olivares, I., Karger, D. N., & Kessler, M. (2018). Assessing species saturation: Conceptual and methodological challenges. *Biological Reviews, 93*, 1874–1890.
Pfennig, D. W., & Pfennig, K. S. (2010). Character displacement and the origins of diversity. *American Naturalist, 176*(Suppl 1), S26–S44.
Potts, R., & Behrensmeyer, A. K. (1992). Late Cenozoic terrestrial ecosystems. In A. K. Behrensmeyer et al. (Eds.), *Terrestrial ecosystems through time: Evolutionary paleoecology of terrestrial plants and animals* (pp. 419–541). University of Chicago Press.
Prevosti, F. J., Forasiepi, A., & Zimicz, N. (2013). The evolution of the Cenozoic terrestrial mammalian predator guild in South America: Competition or replacement? *Journal of Mammalian Evolution, 20*, 3–21.
Rampino, M. R., Haggerty, B. M., & Pagano, T. C. (1997). A unified theory of impact crises and mass extinctions: Quantitative tests. *Annals of the New York Academy of Sciences, 822*, 403–431.
Reznick, D. N., & Ricklefs, R. E. (2009). Darwin's bridge between microevolution and macroevolution. *Nature, 457*, 837–842.
Ricklefs, R. E. (1979). *Ecology* (2nd ed.). Chiron Press.

Sahney, S., Benton, M. J., & Falcon-Lang, H. J. (2010). Rainforest collapse triggered Carboniferous tetrapod diversification in Euramerica. *Geology, 38*, 1079–1082.
Sakamoto, M., Benton, M. J., & Venditti, C. (2016). Dinosaurs in decline tens of millions of years before their final extinction. *Proceedings of the National Academy of Sciences USA. 113*, 5036–5040.
Schluter, D. (2000). *The ecology of adaptive radiation*. Oxford University Press.
Sepkoski, J. J., Jr. (1979). A kinetic model of Phanerozoic taxonomic diversity, II: Early Phanerozoic families and multiple equilibria. *Paleobiology, 5*, 222–252.
Shear, W. A., & Edgecombe, G. D. (2010). The geological record and phylogeny of the Myriapoda. *Arthropod Structure and Development, 39*, 174–190.
Silvestro, D., Antonelli, A., Salamin, N., & Quental, T. B. (2015). The role of clade competition in the diversification of North American canids. *Proceedings of the National Academy of Sciences USA, 112*, 8684–8689.
Simpson, G. G. (1944). *Tempo and mode in evolution*. Columbia University Press.
Simpson, G. G. (1953). *The major features of evolution*. Columbia University Press.
Solé, R. V., Montoya, J. M., & Erwin, D. H. (2002). Recovery after mass extinction: Evolutionary assembly in large-scale biosphere dynamics. *Philosophical Transactions of the Royal Society of London B, 357*, 697–707.
Stanley, S. M. (1999). *Earth system history*. W. H. Freeman and Company.
Strauss, B. (2020). 10 myths about dinosaur extinction. *ThoughtCo*, February 11, 2020, thoughtco.com/myths-about-dinosaur-extinction-1092145
Valkenburgh, B. V. (1999). Major patterns in the history of carnivorous mammals. *Annual Reviews of Earth and Planetary Sciences, 27*, 463–493.
Vermeij, G. J. (2004). Ecological avalanches and the two kinds of extinction. *Evolutionary Ecology Research, 6*, 315–337.
Voje, K. L., Holen, Ø. H., Liow, L. H., & Stenseth, N. C. (2015). The role of biotic forces in driving macroevolution: Beyond the Red Queen. *Proceedings of the Royal Society of London B, 282*, 20150186.
Wahlberg, N., Wheat, C. W., & Pena, C. (2013). Timing and patterns in the taxonomic diversification of Lepidoptera (butterflies and moths). *PLoS One, 8*(11), e80875. https://doi.org/10.1371/journal.pone.0080875
Wang, X., Tedford, R., & Antón, M. (2008). *Dogs: Their fossil relatives and evolutionary history*. Columbia University Press.
Wesley-Hunt, G. D. (2005). The morphological diversification of carnivores in North America. *Paleobiology, 31*, 35–55.
Wilson, E. O., & Hölldobler, B. (2005). Eusociality: Origin and consequences. *Proceedings of the National Academy of Science USA, 102*, 13367–13371.
Wilson, G. P., Evans, A. R., Corfe, I. J., Smits, P. D., Fortelius, M., & Jernvall, J. (2012). Adaptive radiation of multituberculate mammals before the extinction of dinosaurs. *Nature, 483*, 457–460.

Chapter 18
The Origin of Life and Self-Assembly of Modern Nutrient Cycles

The data on the origin of life can hardly be called facts for the sole reason that the number of opinions almost matches the number of researchers. Nevertheless, there is a tendency to think that life first emerged in the ocean some 4–3.8 million years ago. Back then, the atmosphere and the water bodies most likely already contained the materials needed to produce life, such as CO_2, CO, NH_3, CH_4, H_2, N_2, SO_2, H_2S, and N_mO_n (see, e.g. Catling & Zahnle, 2020). They were part of the volcanic gas and substances produced during the reaction of this gas with hydroxyl radicals or under the impact of short-wave radiation.

For quite a long period, since the 1930s, the prevailing view was that, at first, a solution of organic material, the so-called primordial soup or prebiotic broth, had been produced from these inorganic substances by way of chemical polymerisation. Eventually, these organic substances had turned into the first cells or protobionts. The latter were already capable of growth, reproduction, and mutation, so they could evolve. Most probably, they were feeding on the 'broth', so they performed as decomposers. Let us call this hypothesis the heterotrophic origin, although it has been long known as the Oparin-Haldane hypothesis. Many of us may remember from our textbooks the laboratory experiments, carried out by S.L. Miller in 1953, that seem to partially support this hypothesis (small organic molecules were produced from inorganic substances under the impact of spark discharges).

However, this hypothesis and its supporting experiments seem to be losing trust among experts. It yields the leading position to other two hypotheses – those of the 'hot cradle' and 'autotrophic origin'. According to the first one (Corliss et al., 1981), the Earth's climate was much hotter at the dawn of life than now; there might have even been a period when the oceans were literally boiling. Nowadays, similar conditions can probably be found only at hydrothermal vents.

The fundament of the autotrophic origin hypothesis was laid by Wächtershäuser (1990). He calls his method 'retrodiction', i.e. projection of the present situation into the past. In principle, this is nothing else but the already mentioned principle of uniformitarianism or actualism (see Chap. 2). Thus, contrary to his predecessors, Wächtershäuser uses the deductive method that he admits having adopted from K. Popper:

> Most workers in the field were inductivists. They believed that the sum total of the experimental results would tell us all about the prebiotic broth. There is perhaps no other example in the whole of science which violates the principles of Popper's methodology of science

E. Lekevičius, *Biodiversity*, https://doi.org/10.1007/978-3-031-11582-0_18

> more thoroughly (…) It is a perfect example for the consequences of a continued application of the wrong methodology of science. (Wächtershäuser, 1997)

At first, he concluded that the 'prebiotic broth' of Oparin-Haldane, if it had ever existed, was lacking the driving force that could push inorganic and small organic molecules up the thermodynamic slope to produce complex organic polymers. A very different situation must have developed at the hydrothermal vents on the deep ocean floor. They eject CO, CO_2, NH_3, H_2S, and other gases into their environment. Nowadays, that driving force is the reaction of hydrogen sulphite erupting from the depths with the oxygen dissolved in the water. This reaction is exploited by colourless sulphur-oxidising bacteria. They feed the rest of the vent community. A similar non-equilibrium system with a thermodynamic and chemical gradient was necessary for the occurrence of life. Under the primordial terrestrial conditions, iron mono- and disulphide (pyrite) performed the role of oxygen, according to Wächtershäuser:

$$\mathrm{H_2S} + FeS \rightarrow FeS_2 + \mathrm{H_2} + en \tag{18.1}$$

$$CO + \mathrm{H_2S} \rightarrow CH_3 - CO - SCH_3 + en \tag{18.2}$$

where CH_3–CO–SCH_3 is methyl thioester of acetic acid. It can stimulate other reactions of chemical synthesis that would happen spontaneously, without demand for energy, since all these reactions, starting with the first one (1) are exothermic, or to be more precise, exergonic (the change in the Gibbs free energy is negative, $\Delta G < O$). This might be the most important Wächtershäuser's idea: life occurred by way of catalysis (reaction 18.1) and autocatalysis (reaction 18.2 and other synthetic reactions). Thus, according to Wächtershäuser, the origin of life was spontaneous; therefore much faster than expected, self-organisation of chemical substances:

> The theory presents the early evolution of life as a process that begins with chemical necessity and winds up in genetic chance. (Wächtershäuser, 1997)

This did not require electric discharges or the energy of light, as the energy for chemical evolution was supplied by the entrails of the planet. Of course, G. Wächtershäuser then had to face the problem of substance concentrations (which also poses a challenge to the hypothesis of heterotrophic origin). Immediately after production, any organic material would be dispersed in the water column of the ocean. The above-mentioned author proposed the following resolution: chemical reactions that lead towards the appearance of the first organism must have taken place on the surface of iron sulphide and nickel sulphide minerals. They not only retained the reagents and the products of reactions but also served as the primary catalysers.

Some postulates of the autotrophic origin hypothesis have successfully passed the laboratory experiment tests. On the other hand, the author himself constantly reviews his hypothesis (e.g. Wächtershäuser, 2006, 2010). He holds the opinion that primary metabolism had been possible without a genetic apparatus, which evolved later as a means to replicate the smoothly functioning biochemical system. In such case, the first organisms might have been pure chemosynthesisers, or rather chemolithoautotrophs. Here 'chemo-' refers to the energy source of chemical origin, 'litho-' indicates an inorganic electron source, and 'auto' specifies the inorganic carbon source, CO_2 in this case. Many scholars believe (e.g. Raven, 2009; Sleep & Bird, 2007, 2008) that the first organisms on Earth might have included chemolithoautotrophs, namely, methanogens and acetogens, which exist also in our times. Methanogens (attributed to archaea) consume H_2 as the source of energy and electrons and CO_2 as the source of carbon, so they are chemolithoautotrophs. Acetogens (part of bacteria) perform a similar reaction of CO_2 assimilation as the one carried out by methanogens; however, the end product is acetate, rather than methane.

The hypothesis of autotrophic origin was further developed by Russell (2006), Martin (Martin & Russell, 2007), and their collaborates (Martin et al., 2014; Xavier et al. 2020). This is what they stated:

- It is possible that life occurred in some microscopic enclosed compartments if these had supply of appropriate inorganic substances, such as H_2, CO_2, NH_3, etc. These conditions are best met in alkaline hydrothermal vents, which have a porous structure and gases, such as methane and hydrogen, percolating from underground, while the temperature at the mouth does not exceed 90 °C. Each micropore with the diameter of some 0.1 mm is a kind of a cell. The synthesis of organic material could be catalysed by iron or nickel monosulphides (FeS and NiS), which used to cover the inner walls of these cells rendering each of them an ideal microscopic reactor.
- At first, these spontaneous synthetic reactions produced the acetic and formic acids and their esters, methane, pyrophosphate, amino acids nucleotides, and lipids.
- Later, RNA, DNA, and proteins evolved, together with the translation mechanism and heritability. Phospholipids formed the structures similar to cellular membranes, which covered the inner walls of these cells. It is possible that not only pyrophosphate but also ATP was already being produced back then, and the external electrochemical gradient was replaced by the internal proton (H^+) pump. At this point, the 'cells' must have become independent of the vents and were able to leave them.
- Organisms have diverged into bacteria and archaea, which differed in their membrane composition, and the formed cells abandoned the porous structures.

Wächtershäuser (2010) presents somewhat different scenario of the origin of genetic apparatus, cell formation, and divergence scenario. Let us focus on divergence for a moment. All the experience of evolutionary biology accrued from the times of Darwin tells us that the evolution of life commenced with just one species, the so-called last universal common ancestor (LUCA). Let us accept that this might be true. It would be the right time to ask then why this primaeval creature had to produce not only its own copies but also numerous other organisms that were different. The easiest answer would be this: even back then, mutagenesis and, possibly, other forms of heritable variation were already in action. This answer would be even more convincing if it also referred to the factors that bring at least some order into the chaos of genetic variation. Oh yes, there was natural selection which was supposed to eliminate all mutants except for that new one destined to become the next 'wild type'. So, let us add it too – that is exactly the reasoning of Wächtershäuser (2010). However, even this extended answer does not mention any factors that increase diversity instead of reducing it. Therefore, the only option is to supplement the previous answers with yet another one: the 'wild' genotype may sound fine, but there is no such a thing as a universal genotype, since each of them is linked to a specific location and time because of the ever-changing environmental conditions (Chaps. 9 and 10).

So, we already have many different genotypes, and all of them are 'wild type', which is good news. But let us move further. After all, we have only approached the factors that might have been responsible for the occurrence of intraspecific diversity. Our hypothetical life form still remains the same protobiont or a very similar organism, albeit widely dispersed in space. And yes, my dear reader, I hope you might have already guessed it right: to make this line of reasoning at least somewhat trustworthy, the gene pool must not be uniform and common to all these organisms. On the contrary, it is necessary that their genotypes do not mix during reproduction, even when their bearers are settled together. Otherwise, any effort to preserve specialisation, i.e. genotype's adaptation to the local conditions, would be useless. The only solution to that is restriction of gene flow by creating a kind of reproductive isolation. Nature must have done exactly that by producing different clones (genetic lines) devoid of crossbreeding ability and different species. My approach to the question of the origin of species diversity might appear too light-minded, which is reflected also by my writing style, possibly resulting from the fact that I have limited capacity to say something original on this matter. For much more serious analysis of the possible mechanisms of this diversification, see Wiedenbeck and Cohan (2011).

While summarising his experience in studying the phenomenon of biodiversity for over 40 years, one of the most outstanding ecologists of our times (Tilman, 2015) wrote the following:

> I propose that both the evolutionary causes and the ecological consequences of biodiversity share the common origin – unavoidable trade-offs that organisms face when dealing with multiple limiting factors (…) the new taxa had trade-off-based differences that let them excel in ecologically empty sites even though the traits that allowed this also precluded them from performing well in sites already occupied by their progenitors (…) This "universal trade-off hypothesis" thus suggests that intraspecific and interspecific trade-offs are evolutionarily unavoidable.

Loxdale et al. (2011) came up with a similar conclusion:

> Most organisms represent specialized forms that arose as a result of natural selection and genetic drift to occupy distinct ecological niches (…) The major thrust of evolution is towards ecological specialisation as a result of the direct effects of intra- and interspecific competition. Adaptation to new resources lowers such competition and allows survival in new habitats/niches.

These authors interpret the process of life diversification on Earth as a never-ending specialisation. According to the reasonable opinion of Reznick and Ricklefs (2009), this tendency has been predicted by the divergence principle, formulated already by Darwin. This principle stipulates that competition is the most intense between individuals of the same population, while it is somewhat weaker between different, even if closely related, species. Therefore, competition and selection act as certain centrifugal forces that turn minor differences between individuals into larger ones, transform varieties into subspecies, the latter – into different species, and so on, up to classes and phyla (the result is character displacement). According to Darwin (1872, pp. 87–103), 'nature's economy' offers a broad range of professions; therefore there was no need for all the organisms ever produced by variation to concentrate on just one or a few of them.

The evolution of life and specialisation of organisms are inseparable, although this does not mean that being a generalist is a disadvantage in all cases (Ma & Levin, 2006). This regularity is exactly the fundamental cause of diversity, both that of genotypes and of species. Which leads to its functional explanation: diversity increases functional efficiency and stability (see Chap. 11).

By the way, biochemists usually provide a similar explanation when speaking about the origin of diversity in sub-cellular structures. Even before the occurrence of the first cell and biodiversity, a certain diversity of organic compounds must have already existed. Part of this diversity would later have become the components of protobiont metabolism. Since back then – just as nowadays – each of these macromolecules were, presumably, suitable only for a very limited function because of their very nature, protobionts could maintain their viability, growth, and reproduction only by coordinating the activities of numerous macromolecules. However, here is the problem: a successful cellular metabolism is possible only under the condition that some biochemical reactions are isolated in space from certain other reactions which are chemically and thermodynamically incompatible with them. Remove the internal membranes of the cell and cell metabolism will halt or go out of control. That is why variation and selection have literally isolated biochemical processes from one another and produced the first sub-cellular structures and their diversity (Cavalier-Smith, 2002; Martin, 2010; see also Kerfeld et al., 2010).

However, let us return to the diversity of species. We stopped at the idea that gene pools of individual species have segregated from one another. And yet this isolation is clearly not absolute – even the most primitive organisms are capable of exchanging genes (by way of transformation, transduction, or conjugation). Sometimes this exchange is not limited to a single gene pool – a horizontal gene transfer from one species to another is also possible. In short, there is isolation, but it is not absolute. Apparently, gene exchange is not a purely negative phenomenon, it can help adapting to the environment which is difficult to predict. This is exactly the reason why natural selection has produced and preserved these precursors of sexual reproduction up till the present day.

Another problem that the ancient life forms must have encountered was the loss of species autonomy resulting from their specialisation. If the first organisms were consumers, sooner or later they must have died out upon depletion of the primordial 'broth' reserve, because the rate or chemical polymerisation could not have possibly matched the reproduction rate of protobionts (Fig. 18.1).

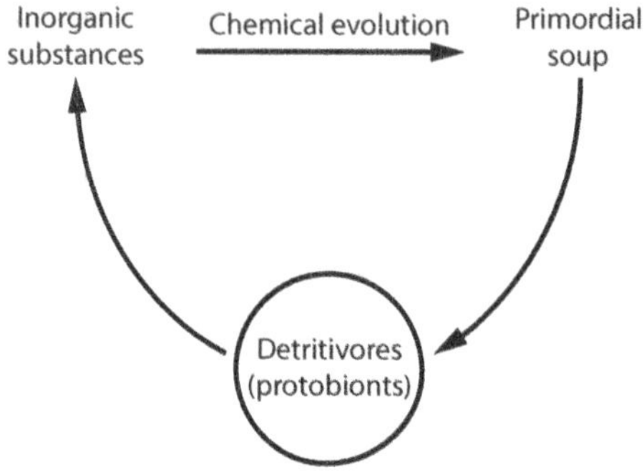

Fig. 18.1 The local material cycle that potentially might have emerged in the case of the heterotrophic origin, although it is unlikely to have been actually achieved because of the incompatibility of rates. (From Lekevičius, 2002))

The alternative where protobionts were autotrophs sounds not much more promising – sooner or later it would have resulted in the depletion of inorganic nutrients. This is understandable, because the expansion of protobionts and increase of their populations would transfer more and more nutrients into organic compounds, making them less and less accessible in the environment. Even more so, since part of the biomass tends to end up as detritus thus locking the nutrients and rendering them unavailable for subsequent reuse.

Some may oppose that all autotrophs without exception perform not only anabolic but also catabolic functions, since they respire or ferment. In other words, they are both producers and decomposers simultaneously. And our opponents would be right – they really are. However, there are at least two buts. Firstly, if autotrophs grow and their population increases its biomass by also generating detritus, it takes more inorganic nutrients from the environment than returns to it, so the nutrient reserve is being drained. Secondly, the possible chemoautotrophs of that period, methanogens and acetogens, excrete as waste into their environment not the nutrients necessary to them (H_2, CO, CO_2, formates), but methane or acetate, i.e. substances that could suit some decomposers but not them themselves. In other words, the missing part was a nutrient cycle, the attribute which provides for an unlimited continuation of life. Only an ecosystem is viable (see Chap. 3). A powerful negative feedback had been created, so either heritable variation together with selection were destined to produce new organisms that would fill in the missing gaps in the material cycle, or else life was doomed to inevitable collapse. Since this collapse has never happened, one may assume that genetic variation and selection produced other bacteria or archaea which could transform that waste into the original reagents for chemosynthesis.

The most likely version, at least as seen nowadays, is that heterotrophs used to appear in places where autotrophs for various reasons were lacking H_2, the likely main source of energy and electrons back then. Martin and Russell (2007) also hint about a possibility that the earliest organisms on Earth were already divided into autotrophs and heterotrophs. In this context, a rather extravagant and original hypothesis of 'ecosystems-first' cannot but be mentioned (Hunding et al., 2006; see also Lekevičius, 2000, 2002; Lekevičius, 2003). Schönheit and co-workers (2016) have yet another vision of the origin of life. In their opinion, the first heterotrophs originated from autotrophs, most probably, soon after life came into being. This reasoning is based on the fact that some modern acetogens, such as *Thermacetigenium phaeum*, function as chemoautotrophs while the partial pressure of H_2 is high; however, when it falls, they become heterotrophs and switch from the synthesis of acetate to its oxidation, i.e. transformation into the nutrients suitable for further chemosynthesis. The authors hence conclude that perhaps very little genetic variation was needed for some autotrophs to become heterotrophs.

If this reasoning has at least some grounds, we should find both autotrophs and heterotrophs among the Earth's oldest organisms. And the period when only one of these groups existed must have been rather short. Unfortunately, there are no direct data that would either support or deny this assumption. Let us test it indirectly then. For instance, we can easily obtain data about the oldest of the presently living organisms, those whose origin dates back some 3.5–3.0 billion years or, preferably, even more. Then we should analyse which type of metabolism could be characteristic of these oldest organisms.

For that purpose, I tapped into the published data (e.g. Basen & Müller, 2017; Battistuzzi et al., 2004; Madigan et al., 2011; Stetter, 2006) about phylogenetic trees compiled by using modern methods (ss rRNA sequencing), as well as the data of other authors on possible metabolism of the oldest organisms. Based on multiple sources, the following list of the possibly oldest genera of archaea and bacteria was produced:

- *Methanococcus*, *Methanothermus*, *Methnotrix*, *Methanopyrus*, *Moorella*, *Thermoanaerobacter*, *Thermacetogenium*, *Pyrodictium*, *Pyrolobus*, *Pyrodium*, *Thermoproteus*, *Archaeoglobus*, *Sulfolobus*, *Desulfurobacterium* (all are anaerobic chemoautotrophs).
- *Pyrobaculum*, *Pyrococcus*, *Thermococus*, *Acetothermus*, *Staphylothermus*, *Desulfurococcus*, *Thermotoga* (all are anaerobic decomposers).

All these organisms are anaerobic thermophiles and hyperthermophiles, i.e. their optimal environment is the one without oxygen and with high temperatures ranging from 45 to 100 °C or more. Nowadays, such temperatures and such organisms are only found at hydrothermal vents; however, in the presumed period of the origin of life, such conditions might have been universal, as mentioned before. Please note that neither photosynthesisers nor aerobic organisms can be found among the possibly oldest life forms.

Thus, we already have an answer, albeit a hypothetical one, to the question what conditions determined the initial stages of diversification and the occurrence of the first nutrient cycles. If we stick to the hypothesis of autotrophic origin, then the appearance of decomposers was presupposed by the occurrence and accumulation of the organic metabolic waste and detritus that served as an adaptive zone. If food is available, the consumers will appear sooner or later. In case of heterotrophic origin, the emergence of autotrophs requires somewhat different explanation. With the depletion of the primordial 'broth', a strong selection pressure to produce any organisms with at least a slightest capacity to extract energy from inorganic material or light must have built up.

Any further elaborations on that topic might be useless since they all would be too speculative, and there is no way to verify them. However, in my understanding, the very conclusion that organisms split into relatively independent species soon after they had come into being sounds rather convincing.

The material presented above suggests that even the primaeval species diversity was already far from chaotic, produced purely by genetic variation; rather, it was organised in a specific way to reflect not only the diversity of abiotic conditions but also the specialisation of species according to the local material cycle and energy flow. In other words, there were some rudiments of ecological communities as coordinated systems. It is quite possible that, with the occurrence of anaerobic chemosynthesisers some 3.8–3.5 billion years ago, ecosystem metabolism was shaped as presented in Fig. 18.2.

Even the most primitive modern organisms can function successfully only if at least a dozen of macro and micro elements are recycled without disruptions (see Chap. 3), which leads to the conclusion that the same elements must have been recycled also in the distant past. Again, this is a private opinion based on the uniformitarianism method.

The reader might be interested to learn that this reasoning is not unprecedented; a similar approach was used by Walker (1980) several decades ago. He also attempted to reconstruct the evolution of ancient ecosystems by applying the principles of ecology, i.e. by way of deduction. First, Walker introduced a few axioms, or premises:

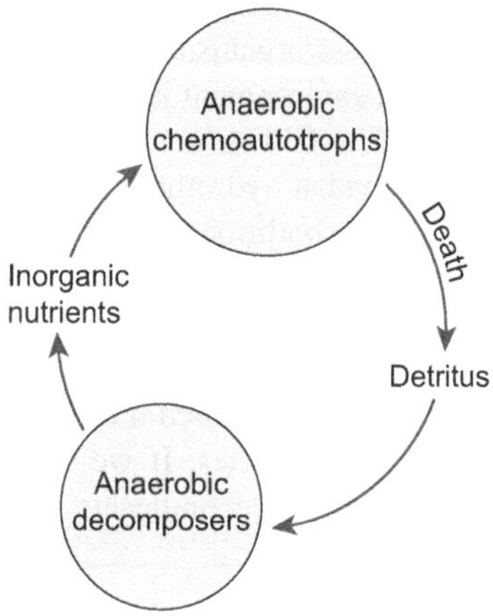

Fig. 18.2 A summarised model of the ecosystems that might have existed some 3.8–3.5 billion years ago, i.e. prior to the occurrence of photosynthesis. (From Lekevičius, 2013)

- Materials circulate in circles, or cycles.
- Material cycles are leaky – some nutrients leave the cycles and become buried in sediments.
- 'Satisfied creatures do not change' – Walker's metaphoric way of saying that organisms do not acquire an ability to use new resources if the old resources are not depleted.
- 'Organisms are greedy' – another metaphor signifying that life tends to increase its total biomass at any cost.

Based on these premises, Walker concluded that evolutionary innovations used to appear as a regular pattern, usually, when older life forms would deplete some important resources (Fig. 18.3). This was inevitably happening from time to time, since 'organisms are greedy'. An ecological crisis would follow, and life would be forced to invent some ways to switch to previously unexploited resources. For instance, the occurrence of the first autotrophs (Walker thinks that these might have been methanogens) was most likely provoked by the circumstances where protobionts, as consumers, have depleted the reserve of organic material produced abiotically. Eventually, however, methanogens were also hit by a crisis once the H_2 reserves had been depleted. Then, and only then ('satisfied creatures do not change'), the emergence of new autotrophs that were not dependent of the availability of H_2 became possible. These most likely were photosynthetic sulphur bacteria, which were using hydrogen sulphide, H_2S, as their e^- source, instead of hydrogen. They must have also been 'greedy' and thus increased the biomass of the biosphere part whereof would leave the cycle in the form of detritus and get deposited into storage; so H_2S has eventually become depleted too. This provoked yet another crisis which was resolved only with the occurrence of cyanobacteria and oxygenic photosynthesis. These bacteria began to use H_2O as an electron source.

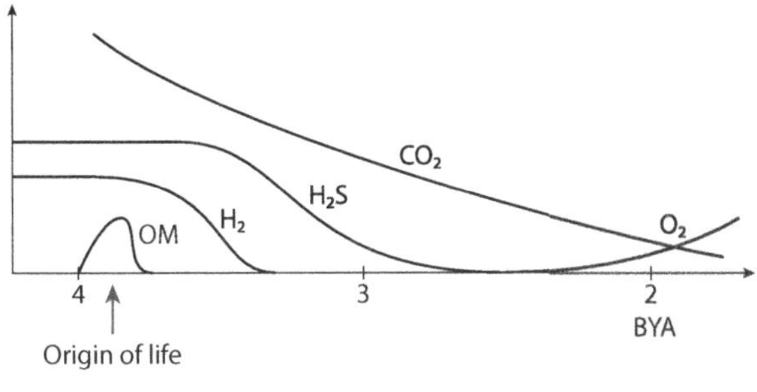

Fig. 18.3 Emergence of new guilds and new nutrient cycles might have been triggered by the depletion of resources resulting in ecological crises (Lekevičius, 2013). The first of them must have hit when the 'primordial broth' had been exhausted, the second one with the depletion of H_2 reserves, while the third one – after the decrease of H_2S volume. The diagram was produced by using some data from Walker (1980). OM – organic material (the product of chemical polymerisation, i.e. the 'primordial broth'). BYA stands for 'billion years ago'

I find this position of Walker (1980) rather well substantiated. It is broadly accepted, for instance, that the total biomass of the biosphere tended to grow, transferring more and more nutrients into the organic compounds, and part of the organic material used to drop out from the cycle in the form of sediments, kerogen (the precursor of oil) and, ultimately, oil (Rasmussen, 2005). However, I am foremost impressed by Walker's courage: he did not hesitate to challenge the mainstream methodology and to demonstrate that it was possible to think and solve problems very differently from the approach traditionally accepted in biology. His deductive strategy must have been so unusual back then, especially in the Western countries, that the author has never used it again, as far as I am aware, as if frightened by his own boldness, which is unfortunate, because his depiction of the early stages of evolution is surprisingly logical and, moreover, alluringly simple.

According to many contemporary authors (e.g. Canfield et al., 2006; Kharecha et al., 2005; Raven, 2009), some 3.8–3.5 billion years ago, photosynthetic (green and purple) sulphur bacteria appeared, just as assumed by Walker. They used light as their energy source, H_2S or Fe^{2+} as the source of electrons, and CO_2 as the carbon supply. One should not assume that no sulphur and iron biogeological cycles had existed until then, since both these elements would have been essential for the functioning of life since its very occurrence. However, the original cycles of these elements were much simpler, since neither H_2S nor Fe^{2+} were used as electron donors at this early stage. See Lekevičius (2002) for more details.

In the discussed period, life had already achieved not only the form of microbial mats but also that of stromatolites. Understandably, the stromatolites of those times have not survived, yet we have their surviving descendants which appeared much later. They are calcified structures resembling a coral reef. In ancient times, the thin upper layer of a stromatolite was most likely produced by green and purple bacteria. Underneath, where the light would not penetrate, a layer of anaerobic decomposers, including the bacteria capable of sulphur, sulphate, and trivalent iron reduction, were situated. These microorganisms would utilise the detritus produced by the photosynthesisers while converting sulphur, sulphates, and trivalent iron into H_2S or Fe^{2+} suitable for new rounds of photosynthesis. Thus, an ancient stromatolite was a relatively independent miniature material cycle (Canfield & Raiswell, 1999; Ohmoto et al., 1993; Schopf, 1992).

In addition to microbial mats, stromatolites, and hydrothermal vents, 3 billion years ago, the biosphere might have already had some open sea communities including photosynthetic plankton (Nisbet 2000). Thus, the overall number of species was gradually increasing.

Cyanobacteria came into being much later, some 2.7 billion years ago (Summons et al. 1999; Nisbet and Sleep 2001), probably after the hydrogen sulphide reserve had been almost depleted, just as Walker assumed. Some time later, they displaced photosynthetic sulphur bacteria from the upper layer of stromatolites and microbial mats. However, the latter, together with the sulphur- and sulphate-reducing bacteria, did not disappear anywhere but rather settled for good underneath the cyanobacteria and their indirect mutualists, aerobic decomposers. Cyanobacteria, apparently, facilitated even broader distribution of these compactly 'packed' ecosystems. One may even assert that they became one of the prevailing life forms from that moment on.

Cyanobacteria might have been the pioneers in using water as the source of electrons and hydrogen. Presumably, that gave them a great advantage, since the reserve of this fuel was practically unlimited. However, cyanobacteria must have encountered one problem. As the result of the oxygenic reaction, they emitted into the environment a new waste product, which was O_2. Oxygen dissolves well in water and is also a strong oxidant, which makes it a poison, unless there is adequate protection. Back then, such protection did not exist: the first bacteria to perform oxygenic photosynthesis undoubtedly were anaerobes, just as all the other organisms of that period. Therefore, some scholars are convinced that there was a mass extinction at that time. There are assumptions that cyanobacteria had developed oxygen resistance first and only later became dependent on it. Eventually, decomposers also acquired aerobic respiration (Eigenbrode & Freeman, 2006), thus becoming satellites and indirect mutualists of cyanobacteria and creating aerobic material cycles. Without these mutualists, cyanobacteria would have played a secondary role in aquatic ecosystems, at most. Instead, some 2.45 billion years ago, they got engaged in adaptive radiation and produced thousands of new species (Schirrmeister et al., 2013).

A short comment seems appropriate here. Cyanobacteria did not pursue any other 'goals' except their own 'agenda' – to use the new more advanced form of photosynthesis for the increase of their total biomass to the maximum. They were successful. Any species would do the same in any times (see Chap. 11). Biomass growth and diversification must have been followed by the accumulation of oxygen and drop of carbon dioxide concentrations in the environment. This outcome could be interpreted as an inevitable by-product of expansion. Naturally, cyanobacteria could not have anticipated that this by-product (O_2) would become useful in the long run, and they would even develop an obligatory dependence on it. Even when cyanobacteria had acquired aerobic respiration, the success was not guaranteed until this catabolic pathway was not taken over also by indirect mutualists. One could hardly find a better illustration of the Gaia hypothesis. On the other hand, I think that this hypothetical example demonstrates quite well the transition from the evolution at species level to the evolution of ecosystems. Selection of ecosystems, as separate units, turns out to be unnecessary.

An interesting and really important detail should be mentioned here. Apparently, life started its activities by acquiring the ability to exploit the best 'fuels' or electron donors, such as H_2, CH_4, and detritus. Only when the reserves of H_2 and CH_4 had been depleted, it switched to lower quality 'fuels' which were less convenient sources of electrons: H_2S, S^o, Fe^{2+}, later to NH^{4+} and Mn^{2+}, and finally towards really poor 'fuel' which was H_2O. In fact, it behaved like a child at a party table who snatches the tastiest goodies first. No wonder that eventually life had to use ever stronger oxidising agents (final electron acceptors): at first, CO_2 and small organic molecules sufficed, later, it had to switch towards SO_4^{2-}, then towards Fe^{3+} and Mn^{4+}, until finally it was forced to involve really strong oxidisers, namely, NO^{3-} and O_2 (Lekevičius, 2011). As we know, this was the right choice. Philosophers, theologians, and deep thinkers in general can only be puzzled by the fact that no matter what Nature undertakes, things eventually work out for the best. However, this topic is beyond the scope of this book. See Nealson (1999) and Nealson and Rye (2004) for more details on these thermodynamic peculiarities.

As a strong oxidiser, the oxygen dissolved in water would enter into chemical reactions with reduced and incompletely oxidised substances, such as H_2, Fe^{2+}, NH_3, H_2S, Mn^{2+}, S^o, or CO. This was a favourable circumstance because during these reactions oxygen would be transformed into less toxic substances (oxides and hydroxides, insoluble iron compounds, sulphates, carbonates, and nitrates). It is thought that some 300–500 million years were needed for oxygen to react with vast quantities of the above-mentioned reduced and incompletely oxidised inorganic substances accumulated in the oceans. Fe^{2+} was especially abundant, and it was transformed into insoluble trivalent iron compounds and hydroxides by way of oxidation. This resulted in large deposits of iron ores (magnetites and haematites) that humans currently use. Consequently, oxygen started accumulating in the atmosphere only after the oceans had been saturated with it, i.e. some 2.2 billion years ago. According to experts, some time later, i.e. about 1.8 billion years ago, the planet was crisscrossed by rivers dyed red with rust. This was caused by the reaction of atmospheric oxygen with iron accumulated on dry land and surfacing from the entrails of the planet (by the way, this process continues even nowadays, although not as intensively). New iron ore deposits were formed, those of ferrihydrites and goethites, which are also exploited by humans to reduce them to elemental iron. This period (2.5–1.7 billion years ago) is quite accurately called the Great Oxidation Event (see Canfield & Teske, 1996; Goldblatt et al., 2006).

Basically, there are reasons to believe that, some 2.5–2.0 billion years ago, the aerobic carbon and oxygen cycle had already attained its present form:

Oxygenic photosynthesis

$$CO_2 + \mathrm{H_2O} \leftrightarrow (CH_2\mathrm{O}) + \mathrm{O_2}$$

Aerobic respiration.

Back then, the direct reaction was performed by cyanobacteria, while the reverse reaction by all aerobic organisms of the biosphere, including cyanobacteria themselves. Presumably, the situation was nearly stationary. However, on geological rather than ecological time scale, such balance might have not been achieved; otherwise there would have been no further growth in oxygen volumes (Grula, 2012; Lenton et al., 2016). The increase of O_2 concentration in the environment might have been possible in any of the following scenarios:

- Overall increase of the biomass
- Increase of the mass of detritus
- Accumulation of the organic fuel reserves (oil, coal, etc.)

It seems that all the three trends manifested themselves both throughout the Archaean and the Proterozoic. Thus, the amount of free and chemically bound oxygen accumulated throughout the entire timespan of evolution must be such that, in the case of a global fire, almost all biomass, detritus, and fossil fuel would be oxidised. This is determined by the stoichiometry of the equations (Cloud & Gibor, 1970). See also Lekevičius (2000, 2002, 2013) for more details.

The earlier-mentioned reduced and incompletely oxidised substances (H_2, Fe^{2+}, NH_3, H_2S, Mn^{2+}, S^o, and CO) have served as food for anaerobic chemoautotrophs and some also for photosynthetic sulphur bacteria, almost since the dawn of life. Therefore, as the reserves of these substances were being depleted due to reaction with oxygen, the ability of the corresponding producers to resist the new competitors was also fading away. So, cyanobacteria of that period may be considered extremely strong competitors. They might have managed to overcome other producers not only because of the switch towards a practically endless source of carbon and electrons but also by contaminating water with oxygen. However, the competitors would seldom be eliminated completely. Some 2.0 billion years ago, waters became stratified into oxygenated (oxic) and anoxic zones (Stüeken et al., 2016), each containing its own sets of producers and decomposers. This helped anaerobic organisms, or at least part of them, to survive until now.

With accumulation of oxygen, water nitrate concentrations started increasing (Garvin et al., 2009; Stüeken et al., 2016), so these effective electron acceptors could be used in respiration and decomposition in the zones where oxygen concentration was low or zero. This led to the occurrence of specific bacteria with nitrate respiration. Cyanobacteria could also utilise nitrates as a supplementary source of nitrogen, in addition to the ammonium ions. Eventually, algae and plants would acquire that skill as well.

The discussed period most likely witnessed the emergence of another guild, that of aerobic chemosynthesisers, or chemolithoautotrophs (Lekevičius, 2002, 2003). Without discussing this, one would have difficulty to understand the assembly of the modern nutrient cycles. As mentioned before, the oxygen produced by cyanobacteria would react with reduced and incompletely oxidised substances. The chemical energy produced by these reactions would be converted into heat. However, such situation could not last long. With the appearance of oxygen in the environment, reduced and incompletely oxidised substances became vacant niches, potential sources of energy and electrons for the future consumers (in this case, the concept of a vacant niche must be interpreted not as reduced and incompletely oxidised substances per se, but each of them coupled with the new oxidising agent, i.e. oxygen). Presumably, even the smallest heritable variations opening access to these free resources would be immediately snatched by natural selection and multiplied. There is no doubt about that – all these niches were taken by newly emerged organisms.

Most likely, that happened somewhere between 2.5 and 2.0 billion years ago (Lekevičius, 2002, 2003).

$$\text{Iron bacteria}: \quad Fe^{2+} + \mathrm{O_2} \rightarrow Fe^{3+} + \text{energy}$$

$$\text{Manganese oxidisers}: \quad Mn^{2+} + \mathrm{O_2} \rightarrow Mn^{4+} + \text{energy}$$

$$\text{Hydrogen oxidisers}: \quad \mathrm{H_2} + \mathrm{O_2} \rightarrow \mathrm{H_2O} + \text{energy}$$

$$\text{Carbon monoxide oxidisers}: \quad CO + \mathrm{O_2} \rightarrow CO_2 + \text{energy}$$

$$\text{Colourless sulphur bacteria}: \quad \mathrm{S^o}, \mathrm{H_2S} + \mathrm{O_2} \rightarrow \mathrm{SO_4}^{2-} + \text{energy}$$

$$\mathrm{H_2S} + \mathrm{NO_3}^{-} \rightarrow \mathrm{SO_4}^{2-}, \mathrm{N_2} + \text{energy}$$

$$\text{Nitrifying bacteria}: \quad NH_4^{+} + \mathrm{O_2} \rightarrow \mathrm{NO_2}^{-} + \text{energy}$$

$$\mathrm{NO_2}^{-} + \mathrm{O_2} \rightarrow \mathrm{NO_3}^{-} + \text{energy}$$

$$\text{Methane oxidisers}: \quad CH_4 + \mathrm{O_2} \rightarrow CO_2 + \text{energy}$$

Nowadays, hundreds of chemolithoautotrophic species, including archaea and bacteria, are found in practically all earthly habitats, different for different species; however, nobody would dare to tell what their diversity was 2 billion years ago.

In short, oxygen accumulation in water made an essential impact on the increase of biodiversity. And again, not for the first time, this diversity was not just a product of chaotic genetic variation; rather, it became organised in specific ways, and its development was subject to certain constraints.

Let us take the evolution of nitrogen cycle as another example. There is some data (Beaumont & Robert, 1999; Falkowski, 1997) that early at the dawn of life waters contained much more nitrogen compounds, especially the ammonium ions, than nowadays (Fig. 18.4a). However, one may reasonably assume that later environmental concentrations of the ammonium ions have declined to a minimum. This must have produced selection pressure which stimulated the emergence of nitrogen-fixing organisms (Fig. 18.4b). Photosynthetic sulphur bacteria might have been the first to gain this ability (Falkowski, 1997; Raven & Yin, 1998; Beaumont & Robert, 1999; but see Weiss et al., 2016). This happened some 3.5–2.75 billion years ago (Stüeken et al., 2016). Much later, some 2.5–2.2 billion years ago, as oxygen was accumulating in the environment (Crowe et al., 2013), atmospheric fixation might have occurred as well (under the impact of lightnings). Approximately at the same time, nitrifying bacteria already existed, too (Garvin et al., 2009; Stüeken et al., 2016). This resulted in nitrites and nitrates accumulating in water (Fig. 18.4c). They must have become vacant niches that provoked the appearance of anammox bacteria and denitrifiers. These organisms exploited oxidised forms of nitrogen as

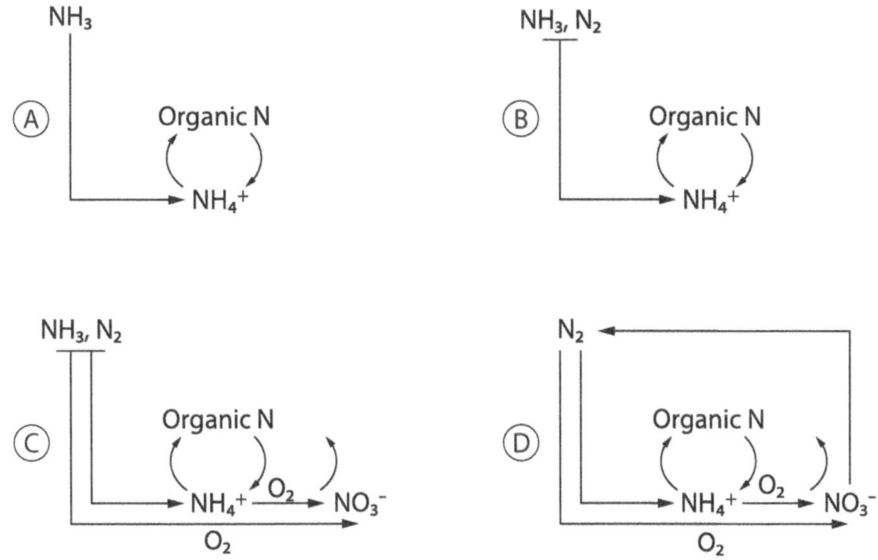

Fig. 18.4 An extremely simplified version of the nitrogen cycle evolution. A, life appeared quite recently; local cycles consisting of assimilation and ammonification are formed; B, ammonium reserves are depleted; biological fixation of nitrogen begins; C, oxygen appears in water and in the atmosphere; nitrates begin to accumulate as the result of lightnings and the activities of nitrifying organisms; nitrates are assimilated by producers; D, denitrification and the anammox reactions occur. (From Lekevičius, 2002)

the irreplaceable oxidisers of ammonium ions (anammox reaction) and detritus (denitrification) under anoxic conditions. Thus, nitrites and nitrates used to be transformed into molecular nitrogen (Fig. 18.4d). The global nitrogen cycle became circular and has remained without essential modifications ever since. Even later, some 500–400 million years ago, a nitrogen cycle was formed also on land according to the already approved template.

To the readers interested in some more detailed information on the possible evolution of material cycles, I could recommend checking my earlier publications (Lekevičius 2000, 2002, 2003, 2009, 2011). Let me emphasise one more idea on this occasion. In the past, we, evolutionary biologists, might have had a reason to focus a lot on the complicated questions raised by our opponents, e.g. how random genetic variation could have produced such sophisticated and advanced organs as the eye, the bird wing, or the brain. While this attention is understood, it is much more difficult to explain the fact that, for so many years, neither we (with some scarce exceptions), nor our opponents posed the question how almost equally complex and advanced structures that we call nutrient cycles have come into being. Whereas the nitrogen cycle may be relatively simple, the same cannot be said about some others, for instance, the sulphur cycle. The scheme of the sulphur cycle resembles to me the patterns on the wing of a tropical butterfly. However, contrary to the latter, these patterns are not directly determined by genes, they do not even belong

Table 18.1 Certain vacant niches in the Precambrian and the organisms that occupied them. Please note that some of the niches were pre-existing while others were created by organisms themselves. According to Lekevičius (2009), with modifications

BYA	Description of vacant niches/adaptive zone	Occupying organisms
3.8(?)	H_2, Fe^{2+}, Mn^{2+}, and H_2S serve as the sources of energy and electrons; CO_2 as the donor of C	Anaerobic chemolithoautotrophs
3.8(?)	Detritus serves as the donor of energy, electrons, and C; organic material, Fe^{3+}, Mn^{4+}, S^o, and SO_4^{2-} act as the ultimate acceptors of electrons	Iron, manganese, sulphur- and sulphate-reducing bacteria
3.5–2.7	Light serves as the energy source; Fe^{2+}, H_2S / H_2O as the donors of electrons; CO_2 as the source of C	Green and purple sulphur bacteria, cyanobacteria
2.5–2.0	H_2, Fe^{2+}, Mn^{2+}, H_2S, CO, NH_3 (NH^{4+}) as the sources of energy and electrons. O_2 as the ultimate electron acceptor	Aerobic chemolithoautotrophs
2.5–2.0	Detritus as the donor of energy, electrons, and C. NO_3^- as the ultimate electron acceptor	Denitrifying bacteria
2.5–2.0	Detritus as the donor of energy, electrons, and C. O_2 as the ultimate electron acceptor	Aerobic decomposers

to a particular species, and yet they are surprisingly stable, transferred from generation to generation, from year to year, even from one geological period to another. And no genetic variation has been able to change or disrupt them for any significant period.

A general picture of the increase of guild and species diversity in the Precambrian, and the underlying reasons thereof, can be obtained from Table 18.1.

References

Basen, M., & Müller, V. (2017). "Hot" acetogenesis. *Extremophiles, 21*, 15–26.

Battistuzzi, F. U., Feijao, A., & Hedges, S. B. (2004). A genomic timescale of prokaryote evolution: Insights into the origin of methanogenesis, phototrophy, and the colonization of land. *BMC Evolutionary Biology, 4*, 44. https://doi.org/10.1186/1471-2148-4-44

Beaumont, V., & Robert, F. (1999). Nitrogen isotope ratios of kerogens in Precambrian cherts: A record of the evolution of atmosphere chemistry? *Precambrian Research, 96*, 63–82.

Canfield, D. E., & Raiswell, R. (1999). The evolution of the sulfur cycle. *American Journal of Science, 299*, 697–723.

Canfield, D. E., & Teske, A. (1996). Late Proterozoic rise in atmospheric oxygen concentration inferred from phylogenetic and sulfur-isotope studies. *Nature, 382*, 127–132.

Canfield, D. E., Rosing, M. T., & Bjerrum, C. (2006). Early anaerobic metabolisms. Philosophical transactions of the Royal Society of London. *Series B, 361*, 1819–1836.

Catling, D. C., & Zahnle, K. J. (2020). The Archean atmosphere. Science. *Advances, 6*, eaax1420.

Cavalier-Smith, T. (2002). The phagotrophic origin of eukaryotes and phylogenetic classification of protozoa. *International Journal of Systematic and Evolutionary Microbiology, 52*, 297–354.

Cloud, P., & Gibor, A. (1970). The oxygen cycle. *Scientific American, 223*, 111–123.

Corliss, J. B., Baross, J. A., & Hoffman, S. E. (1981). A hypothesis concerning the relationship between submarine hot springs and the origin of life on Earth. In *Proceedings 26th International Geological Congress* (pp. 59–69). Geology of Oceans Symposium. , July 7–17, 1980.

Crowe, S., Døssing, L., Beukes, N., Bau, M., Kruger, S. J., et al. (2013). Atmospheric oxygenation three billion years ago. *Nature, 501*, 535–538.

Darwin, C. (1872). *The origin of species by means of natural selection* (6th ed.). John Murray. (Printed in 1998 by Cox & Wyman. Reading, England).

Eigenbrode, J. L., & Freeman, K. H. (2006). Late Archean rise of aerobic microbial ecosystems. *Proceedings of the National Academy of Sciences USA., 103*, 15759–15764.

Falkowski, P. G. (1997). Evolution of the nitrogen cycle and its influence on the biological sequestration of CO_2 in the ocean. *Nature, 387*, 272–275.

Garvin, J., Buick, R., Anbar, A. D., Arnold, G. L., & Kaufman, A. J. (2009). Isotopic evidence for an aerobic nitrogen cycle in the latest Archean. *Science, 323*, 1045–1047.

Goldblatt, C., Lenton, T. M., & Watson, A. J. (2006). Bistability of atmospheric oxygen and the great oxidation. *Nature, 443*, 683–686.

Grula, J. W. (2012). Rethinking the paleoproterozoic great oxidation event. *Abiological Perspective., arXiv*, 1203.6701. (astro-ph.EP).

Hunding, A., Kepes, F., Lancet, D., Minsky, A., Norris, V., et al. (2006). Compositional complementarity and prebiotic ecology in the origin of life. *BioEssays, 28*, 399–412.

Kerfeld, C. A., Heinhorst, S., & Cannon, G. C. (2010). Bacterial microcompartments. *Annual Review of Microbiology, 64*, 391–408.

Kharecha, P., Kasting, J., & Siefert, J. (2005). A coupled atmosphere-ecosystem model of the early Archean earth. *Geobiology, 3*, 53–76.

Lekevičius, E. (2000). *Only an ecosystem is alive: A somewhat nontraditional approach to the evolution of life*. Vilnius University. (in Lithuanian).

Lekevičius, E. (2002). *The origin of ecosystems by means of natural selection*. Institute of Ecology. (Vilnius).

Lekevičius, E. (2009). Vacant niches in nature, ecology, and evolutionary theory: A mini-review. *Ekologija (Vilnius), 55*, 165–174.

Lekevičius, E. (2011). Ecological Darwinism or preliminary answers to some crucial though seldom asked questions. In L. E. Grinin, A. V. Korotaev, R. L. Carneiro, & F. Spier (Eds.), *Evolution: Cosmic, biological, and social* (pp. 101–121). Uchitel Publishing House.

Lekevičius, E. (2013). *Ecology: From individuals to the biosphere*. Vilnius University. (in Lithuanian).

Lekevičius, E., & (Лекявичюс Э.). (2003). Ecosystem evolution: The major stages and possible mechanisms. *Journal of General Biology (Moscow), 64*(5), 371–388. (in Russian).

Lenton, T. M., Dahl, T. W., Daines, S. J., Mills, B. J., Ozaki, K., et al. (2016). Earliest land plants created modern levels of atmospheric oxygen. *Proceedings of the National Academy of Sciences USA., 113*, 9704–9709.

Loxdale, H. D., Lushai, G., & Harvey, J. A. (2011). The evolutionary improbability of 'generalism' in nature, with special reference to insects. *Biological Journal of the Linnean Society, 103*, 1–18.

Ma, J., & Levin, S. A. (2006). The evolution of resource adaptation: How generalist and specialist consumers evolve. *Bulletin of Mathematical Biology, 68*, 1111–1123.

Madigan, M. T., Martinko, J. M., Stahl, D., & Clark, D. P. (2011). *Brock biology of microorganisms* (13th ed.). Benjamin Cummings.

Martin, W. (2010). Evolutionary origins of metabolic compartmentalization in eukaryotes. *Philosophical Transactions of the Royal Society B, 365*, 847–855.

Martin, W., & Russell, M. J. (2007). On the origin of biochemistry at an alkaline hydrothermal vent. *Philosophical Transactions of the Royal Society of London B, 362*, 1887–1926.

Martin, W. F., Sousa, F. L., & Lane, N. (2014). Energy at life's origin. *Science, 344*, 1092–1093.

Nealson, K. H. (1999). Post-Viking microbiology: New approaches, new data, new insights. *Origins of Life and Evolution of the Biosphere, 29*, 73–93.

Nealson, K. H., & Rye, R. (2004). Evolution of metabolism. In W. H. Schlesinger (Ed.), *Biogeochemistry. Treatise on geochemistry* (Vol. 8, pp. 41–61). Elsevier Pergamon.

Nisbet, E. (2000). The realms of Archaean life. *Nature, 405*, 625–626.

Nisbet, E. G., & Sleep, N. H. (2001). The habitat and nature of early life. *Nature, 409*, 1083–1091.

Ohmoto, H., Kakegawa, T., & Lowe, D. R. (1993). 3,4 – Billion – Year – Old biogenic pyrites from Barberton, South Africa: Sulphur isotope evidence. *Science, 262*, 555–557.

Rasmussen, B. (2005). Evidence for pervasive petroleum generation and migration in 3,2 and 2,63 Ga shales. *Geology, 33*, 497–500.

Raven, J. A. (2009). Contributions of anoxygenic and oxygenic phototrophy and chemolithotrophy to carbon and oxygen fluxes in aquatic environments. *Aquatic Microbial Ecology*, 1–16. https://doi.org/10.3354/ame01315

Raven, J. A., & Yin, Z.-H. (1998). The past, present and future of nitrogenous compounds in the atmosphere, and their interactions with plants. *The New Phytologist, 139*, 205–219.

Reznick, D. N., & Ricklefs, R. E. (2009). Darwin's bridge between microevolution and macroevolution. *Nature, 457*, 837–842.

Russell, M. (2006). First life. *American Scientist, 94*, 32–39.

Schirrmeister, B. E., Vosb, J. M. d., Antonellic, A., & Bagheri, H. C. (2013). Evolution of multicellularity coincided with increased diversification of cyanobacteria and the great oxidation event. *Proceedings of the National Academy of Sciences USA., 110*, 1791–1796.

Schönheit, P., Buckel, W., & Martin, W. F. (2016). On the origin of heterotrophy. *Trends in Microbiology, 24*, 12–25.

Schopf, J. W. (1992). The oldest fossils and what they mean. In J. W. Schopf (Ed.), *Major events in the history of life* (pp. 29–63). Jones and Bartlett.

Sleep, N. H., & Bird, D. K. (2007). Niches of the pre-photosynthetic biosphere and geological preservation of earth's earliest ecology. *Geobiology, 5*, 101–117.

Sleep, N. H., & Bird, D. K. (2008). Evolutionary ecology during the rise of dioxygen in the earth's earliest biosphere. *Philosophical Transactions of the Royal Society of London B, 363*, 2651–2664.

Stetter, K. O. (2006). Hyperthermophils in the history of life. *Philosophical Transactions of the Royal Society B, 361*, 1837–1843.

Stüeken, E. E., Kipp, M. A., Koehler, M. C., & Buick, R. (2016). The evolution of Earth's biogeochemical nitrogen cycle. *Earth-Science Reviews, 160*, 220–239.

Summons, R. E., Jahnke, L. L., Hope, J. M., & Logan, G. A. (1999). 2-Methylhopanoids as biomarkers for cyanobacterial oxygenic photosynthesis. *Nature, 400*, 554–560.

Tilman, D. (2015). Biodiversity: From evolutionary origins to ecosystem functioning. *Contributions to Science., 11*, 11–20.

Wächtershäuser, G. (1990). Evolution of the first metabolic cycles. *Proceedings of the National Academy of Sciences USA., 87*, 200–204.

Wächtershäuser, G. (1997). The origin of life and its methodological challenge. *Journal of Theoretical Biology, 187*, 483–494.

Wächtershäuser, G. (2006). From volcanic origin of chemoautotrophic life to bacteria, archaea and Eukarya. *Philosophical Transactions of the Royal Society B, 361*, 1787–1808.

Wächtershäuser, G. (2010). Chemoautotrophic origin of life: The iron-sulfur world hypothesis. In L. L. Barton et al. (Eds.), *Geomicrobiology: Molecular and environmental perspective* (pp. 1–35). Springer.

Walker, J. C. G. (1980). Atmospheric constraints on the evolution of metabolism. *Origins of Life, 10*, 93–104.

Weiss, M. C., Sousa, F. L., Mrnjavac, N., Neukirchen, S., Roettger, M., et al. (2016). The physiology and habitat of the last universal common ancestor. *Nature Microbiology, 1*, 16116.

Wiedenbeck, J., & Cohan, F. M. (2011). Origins of bacterial diversity through horizontal genetic trasfer and adaptation to new ecological niches. *FEMS Microbiology Reviews, 35*, 957–976.

Xavier, J. C., Hordijk, W., Kauffman, S., Steel, M., & Martin, W. F. (2020). Autocatalytic chemical networks at the origin of metabolism. *Proceedings of the Royal Society B, 287*, 20192377.

Chapter 19
Appearance of Modern Ecological Pyramids: Summing Up

It is considered that for at least 2 billion years life had managed with two out of three 'functional kingdoms' – producers and decomposers, as consumers (strictly speaking, macroconsumers, or biophages – see Chap. 3) did not exist. Some people already rushed to label this period 'the Garden of Eden', the implication being of a peaceful coexistence where nobody killed anybody. In other words, back then, the biomass was still a giant adaptive zone, an array of unoccupied niches waiting for their consumers.

Ecologists understand consumers as herbivores, carnivores, and parasites. These terms might be subject to a different interpretation, so let me clarify: hereafter, I am going to call primary consumers, or herbivores, all organisms foraging on decomposers or producers; the ones feeding on the latter would be referred to (as they typically are in ecology) primary carnivores; these are followed by secondary carnivores, etc. My understanding of parasites comprises those organisms which, as a rule, are smaller than their host, exploit it, and cause a certain damage although not necessarily kill it directly.

Most evolutionary biologists agree that consumerism is linked with the occurrence of unicellular eukaryotes (Cavalier-Smith 2002; Knoll et al. 2006; Davidov and Jurkevitch 2009). It is presumed that a cell nucleus might have evolved from a fold of the plasma membrane. According to the Endosymbiosis theory, mitochondria originate from some aerobic prokaryotic consumer which at first was an intracellular parasite. Eventually, this parasite lost its autonomy and became an organelle performing respiratory functions. Other hosts might have applied the same approach not only towards an incorporated consumer but also towards a smaller cyanobacterium which eventually had lost many genes and become a chloroplast. In this way, the first algae might have appeared, in the beginning – as unicellular forms. Here I just presented, with some simplifications, a theory that has by now become much more refined and represents a broad field of studies where at least one new hypothesis is produced almost annually.

There is no common agreement on when the first eukaryotes appeared. Those using the methods of molecular phylogenetics claim that they have a proof of this having happened some 2 billion years ago (for an overview, see Hedges et al. 2015), whereas the fossil data usually suggest a later date – some 1.6–1.4 billion years ago (Vidal 1984; Javaux et al. 2004). Admittedly, palaeontologists also present some data regarding molecular traces of life embedded in rocks, or biomarkers, which

E. Lekevičius, *Biodiversity*, https://doi.org/10.1007/978-3-031-11582-0_19

seem to imply that eukaryotes appeared much earlier, possibly even more than 2 billion years ago.

A common agreement might be difficult to arrive at for yet another reason. It is assumed that oxygen concentrations in the atmosphere and water bodies could be much lower in the period between 2 and 1 billion years ago (although see Blamey et al. 2016), compared to the present levels, and it is well known that almost all eukaryotes are obligatory aerobes. Therefore, numerous hypotheses have been proposed suggesting that the early eukaryotes did not need much oxygen as they were facultative aerobes (e.g. Martin et al. 2015). Many authors consider that they might have been protists related to modern amoebas or primitive flagellates, and it seems highly probable that they also were the first primary consumers (see, e.g. the overview by Javaux 2007). I find it all quite convincing. One could add that some data of molecular phylogenetics seem to indicate that amoebas (Class Amoebozoa) occurred 1.47 billion years ago (Hedges et al. 2015).

Unicellular algae might have evolved somewhat later. They include the famous, albeit still mysterious, acritarchs, which were abundant 1.9–0.6 billion years ago and then disappeared (Schopf 1992; Bengtson 2002). Their cells were large and of various shapes, cell walls were thick, and many of them had various spikes and projections. In the rocks of the Cryogenian Period (850–635 million years ago), palaeontologists discovered a considerable diversity of other microfossils, which were labelled scale fossils. They established (see the overview by Porter 2011) that these were most likely unicellular algae as well.

From the very appearance of protists, in many cases their cell sizes had been steadily increasing (Vidal 1984; Schopf 1992; Bengtson 2002). They became especially large 1.1–0.7 billion years ago when they reached 1 cm in the diameter. As we know, prokaryote cells are ten or more times smaller. This difference was of great importance, since the primary consumers and predators would gain an advantage by outgrowing their prey; on the other hand, the prey species also benefited from a bigger body size which facilitated their defence from predation – see Nooijer et al. (2009) for theoretical justification. Another well-known defence tactic is the acquisition of various external projections, spikes, armour, or thicker external cover. It turns out that this tendency existed in the discussed period as well (Bengtson 2002). Finally, there was a possibility to escape predation by fleeing or hiding. This trend would become more conspicuous somewhat later, especially in the Cambrian.

It is difficult to draw overarching conclusions from this rather confusing material, but one could state, with some degree of certainty, that three trophic levels already existed some 1.4–1.2 billion years ago. The first of them was represented by algae, archaea, and bacteria, while the second and the third one belonged to protozoans. In terms of their trophic niches, the modern genera of *Euglena*, *Peranema*, and *Amoeba* should be rather close to the protozoans of that period. They include the organisms that I would describe as primary consumers and secondary consumers, or primary predators. Besides, the protozoans of that time might have also included some decomposers, as is the case in our times. The presence of parasites in the discussed period seems also highly probable (although parasitologists, apparently, tend to believe that the objects of their studies emerged much later – e.g.

Bryant 1994). Unfortunately, the available fossil records of that time offer only modest information. They are limited to microfossils encased in mineralised shells dating back 770–742 million years, which are called vase-shaped microfossils. They are considered to be protists close to the modern euglyphid amoebas (Porter 2011); some of them might have been non-specialised predators foraging on bacteria, algae, and other smaller protists. The author considers these fossils to be proof that unicellular predators, capable of engulfing or penetrating their prey, already existed in the specified period. This would explain the occurrence of various protrusions, shells, and even spines in this evolutionary period. It is quite possible, further elaborates the author, that predators had evolved even earlier, when neither they themselves nor their prey had shells or scales. They had been feeding exclusively on bacteria, but eventually, with the occurrence of algae, which were much larger than bacteria, some of them were compelled to increase their body mass and switch to that new type of prey. However, there is no fossil record preserved as these predators did not have any exoskeleton and thus were unlikely to fossilise. Algae had developed shells, and primary consumers (e.g. amoebae) had to follow the suit, because they themselves had become a target for larger predators (Ibid.).

Let us move on towards multicellular organisms. They might have occurred as early as in the Mesoproterozoic, i.e. over 1 billion years ago, and they were algae (see the overview by De Clerck et al. 2012). For a long time, it was considered that multicellular animals evolved only in the second half of the Ediacaran (635–545 million years ago), while their adaptive radiation started in the Cambrian (545–488 million years ago). Presumably, almost all phyla of the modern fauna had formed merely within the first 25 million years of that period. Hence the reference to the 'Cambrian explosion', the term meant to emphasise the incredible speed of adaptive radiation that produced hundreds or even thousands of new species in relatively short time span. Later, a few decades ago, the discovery of new fossils pushed the date somewhat back in time, and the 25 million years became almost 50 million (Erwin et al. 2011), however, even that is not a lot to evolutionary biologists. Molecular data (Erwin et al. 2011; Wheat and Wahlberg 2013) also seem to indicate that organisms close to the modern placozoans, sponges, velvet worms, priapulids, nematodes, nemerteans, brachiopods, molluscs, rotifers, coelenterates, ringed worms, and arthropods had already existed in the Ediacaran.

Does that mean that the 'explosion' began not in the Cambrian but much earlier, in the Ediacaran Period? If so, then where are the fossils? There is no clear answer, although a common response to criticism is that multicellular organisms of that time did not have mineralized or chitin skeletons yet; therefore, they would rarely fossilize.

As if that were not enough, still other works in molecular phylogenetics state that coelenterates occurred 960 million years ago, echinoderms 767 million years ago, and sponges 733 million years ago (Hedges et al. 2015; see also Wray et al. 1996; Adoutte et al. 2000). Besides, already in the year 2000, J. Butterfield described the famous multicellular fossil of *Bangiomorpha pubescens*, dated back 1.2 billion years, which was surprisingly similar to the modern red algae *Bangia*. In short, there are more questions than answers.

Presumably, bacteria, unicellular algae, and protozoans must have provoked the appearance of large filter feeders and multicellular predators, which would exploit them. But then why was multicellularity so much delayed? Biologists tend to think that, compared to monocellular protists, a much higher concentration of O_2 was necessary for multicellular life, reaching at least one-tenth of the present levels. It is well known that, even at the present concentration, the bodies of many multicellular organisms, e.g. flatworms, cannot be thicker than 1 mm, because diffusion is only able to supply oxygen to the tissues within 0.5 mm from the skin surface, and these worms do not have blood vessels and respiratory system. The atmospheric oxygen concentration approached the modern levels only around 0.5 billion years ago, while earlier, for instance, 1 billion years ago, it was much lower (although see Blamey et al. 2016). According to some authors, this circumstance was the key factor in the occurrence and spread of multicellular organisms during that period. Only time will tell whether this conclusion is correct.

One may think, a priori, that there were also other reasons why the evolution of multicellulars could not proceed fast. Originating from the colonial lifestyle, it needed many million years to transform into mutualism based on cell (tissue) specialisation, obligatory interdependence, and strict functional coordination of the entire organism. Heritable variation and selection had to invent unprecedented ways to ensure transportation of energy, materials, and information from cell to cell.

Anyway, let us return to the ecology of the first multicellular life forms. The common ancestor of multicellulars must have been similar to the modern placozoans or sponges (Erwin et al. 2011); the latter filter water, thus retrieving zooplankton, phytoplankton, and detritus particles. According to these authors, the first coelenterates joined them some 690 million years ago, in the late Cryogenian. They captured somewhat larger food. Yet it seems that even later, in the Ediacaran, there were no predators foraging on other animals (Xiao and Laflamme 2008). So, there were hardly more than three trophic levels. This means that many vacant niches were available in the form of unexploited prey, suitable for future large predators.

We should not be surprised to see them appearing some time later, probably after a successful survival through the intermediary phase of cannibalism. No spare food can remain as such for a long time. Bengston (2002; see also Vannier 2012) presents the following picture of the ecological pyramid from the early Cambrian (545–488 million years ago), which followed the Ediacaran:

- The first level: Detritus, dissolved organic matter, cyanobacteria, and algae.
- The second level: Bacteria, sponges, molluscs, brachiopods, arthropods, priapulids, worms, echinoderms.
- The third level: Priapulids, arthropods, coelenterates, ctenophores, worms, anomalocaridids.
- The fourth level: Ctenophores, coelenterates, anomalocaridids.

All these groups of organisms (except anomalocaridids) have survived until modern times. Anomalocaridids were still small in the early Cambrian, but eventually these predators would reach indeed impressive dimensions, before being displaced by cephalopods in the subsequent Ordovician Period. There is no evidence that any organisms that would feed on multicellular algae and could be called typical herbivores already existed in the Cambrian; they are thought to have evolved only at the end of the Palaeozoic (Vermeij and Lindberg 2000). Thus, multicellular algae, in their living form, had managed to remain an unoccupied adaptive zone for 700 million years, right? This is hard to believe, but these palaeontologists assume so. Of course, there are also alternative opinions that herbivores appeared in as early as the Cambrian (LoDuca et al. 2017), so their tardiness is cut short from 700 to 400 million years or slightly more. Still, even that seems a lot. Some scientists (e.g. Bengtson 2002) believe that this period was shorter yet: stromatolites began to decline as early as in the Proterozoic, i.e. in the Precambrian, and it is quite possible that some grazers contributed to that. Their identity, however, remains unknown.

There is a widespread opinion that the 'Cambrian explosion' was caused by the appearance of relatively large predators early in this period. Arguably, they might have provoked diversification among their prey species encased in exoskeletons. However, just as some other authors (e.g. Zhang et al. 2014), I consider that in this case the outcome might be mistaken for the cause. I tend to think that first, still in the Ediacaran, the future prey underwent diversification and they did not have mineralised shells since there were no effective predators, as mentioned before. The latter appeared later, in the early Cambrian, and might well promoted the occurrence of armoured prey by starting 'the arms race'.

Even though I have some remarks on the S. Bengtson's pyramid presented above – especially because it does not entirely match the picture presented by Erwin with the co-authors (2011) and Dunne with his colleagues (2008), let us assume that it reflects the reality pretty well. In that case, we shall have to admit that it is at least two or three levels short of the pyramid characteristic to the modern oceans. Even in the late Cambrian, this shortage was compensated only partly. Ecological pyramids would acquire the modern shape only by the end of the next Period (Lekevičius 2002), the Ordovician (488–444 million years ago), with the upswing of cephalopods, which already included species up to several metres long. Ever since, the availability of vacant niches dropped sharply, which might also be reflected by the stabilisation of the number of taxons observed from the Silurian onwards (Sepkoski 1979; Alroy et al. 2008). Large aquatic organisms that would evolve later, such as fishes, would have no other means to establish themselves but by displacing part of their competitors. This would have a negative effect on the diversity of not just cephalopods, but also chelicerates and trilobites. Of course, the winners would eventually produce a similar or even greater number of species. I already described a comparable situation on land regarding the clash between amphibians and reptiles in the Permian (Chap. 17).

So, the current data do not contradict the opinion expressed already by Hutchinson (1959) that evolutionary assemblage of aquatic ecosystems proceeded with acceleration. He wrote:

> Early during evolution the main process from the standpoint of community structure was the filling of all the niche space potentially available for producer and decomposer organisms and for herbivorous animals. As the latter, and still more as carnivorous animals began to appear, the persistence of more stable communities would imply splitting of niches

> previously occupied by single species as the communities became more diverse (…) As this process continued one would expect the overall rate of evolution to have increased, as the increasing diversity increased the probability of the existence of species preadapted to new and unusual niches (…) We may perhaps therefore see in the process of evolution an increase in diversity at an increasing rate till the early Palaeozoic, by which time the familiar types of community structure were established.

A special attention should be given to the inhabitation of freshwater bodies. One may assume (Gray 1988; Sanchez-Baracaldo et al. 2017) that the first colonists were microorganisms and unicellular algae, which had arrived in these habitats prior to the occurrence of plants. The microorganisms most likely included also decomposers feeding on detritus washed down from the land habitats. As seen from the fossils, animals began to appear in these habitats as early as the Cambrian (Gray 1988). There is no doubt that in the Devonian lakes and rivers already hosted significant and ever-increasing diversity of life. Many of them, especially insects, might have evolved from the precursors already dwelling on land, yet another source of origin was the salty waters (Miller and Labandeira 2002; Park and Gierlowski-Kordesch 2007). Unfortunately, we can tell very little about the first pioneers, the adaptive zones, and vacant niches that they produced. One could assume a priori that formation of these ecosystems was hardly significantly different from that in other aquatic habitats (see also Minter et al. 2016).

But let us move towards the conclusion. According to the currently prevailing hypothesis of the autotrophic origin, producers were the first ones to emerge in an aquatic environment. A little later, they produced decomposers. Much later, in perhaps over 2 billion years, the first consumers evolved from some producers or decomposers. They, in turn, most likely produced representatives of the third trophic level, the primary predators, which then automatically became vacant niches and stimulated the evolution of next-level predators and so on (Fig. 19.1). This is a natural sequence that should not surprise us since it is typical of both the succession and evolution on land (Chapters 14, 15, 16 and 17).

However, the rates of assembly of the nutrient cycles and ecological pyramids were very different on land and in water. In water, this process had lasted for over 3 billion years, while on land only a few hundred million years were enough. Why so? The principal answer seems to be quite obvious (Lekevičius 2018): contrary to the evolution on land, aquatic life had to emerge from non-living matter, and afterwards it had to manage with the stuff at its disposal, without any readily available raw material. The terrestrial life, on the other hand, emerged a billion years ago or even earlier from the 'prefabrications' already abundant in water, which required only limited genetic changes to move over to the dry land (see Tables 16.1 and 17.1). These organisms created the first nutrient cycles in the new environment. Later, some small invertebrates also moved from water to land and formed the second or even third trophic level. Again, not much variation was necessary; therefore, the transition was neither complicated nor too long. Furthermore, simultaneously with the transfer of 'prefabrications', in situ evolution was rather swift: typical ersatz at first, soon terrestrial life forms diversified and became much more adapted; besides,

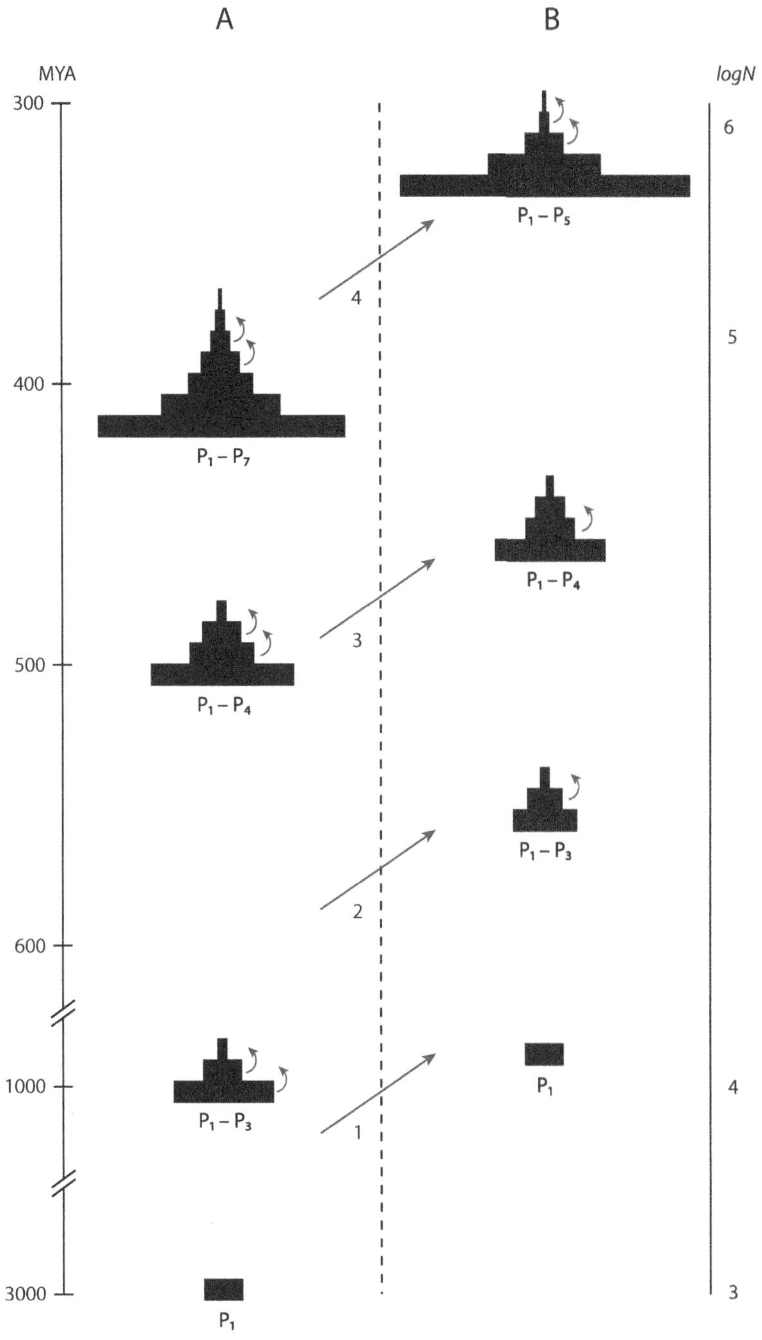

Fig. 19.1 The likely assemblage stages of aquatic and terrestrial ecological pyramids. *MYA*, million years ago; *N*, approximate number of species in the biosphere; $P_1 - P_n$, the levels of production pyramid. The long arrows represent the phases in the terrestrial expansion of aquatic organisms ('raw materials'): the presumed appearance of unicellular producers and decomposers in wetlands and coastal areas (1); the emergence of primary consumers and primary predators on land (2); the subsequent diversification of consumers and the emergence of secondary (3) and tertiary (4) predators on land. The short arrows pointing from the lower to the higher trophic levels represent in situ production of the new level from the organisms in the lower level

in the absence of large predators, they started producing them themselves after passing through the phase of cannibalism first (Fig. 19.1; see also Fig. 15.1).

This scenario is especially probable, if the transformation of an existing aquatic predator into a terrestrial form was more complicated than the emergence of predatory forms of the already terrestrial myriapods, acarines, or nematodes.

After all, even the form, structure, or arrangement that biodiversity tends to attain eventually, as it develops, is almost independent of the medium where this development takes place, be it water or land. As we have hopefully learned, it is barely dependent on the mechanism of community self-assembly either, regardless whether it is ecological, evolutionary, or a combination of these. In short, the phenomenon of functional convergence of ecosystems is strikingly universal. Evolution and succession are the two forms of self-organisation of life that present similar solutions again and again, as we had a chance to find out. It is hardly surprising, since the same laws of thermodynamics and the same types of biological constraints apply to both terrestrial and aquatic environment, both in the past and in the present.

Pianka (1978) once presented an idea that many would find quite unusual: biologists should start thinking about the possible compilation of a 'periodic table of niches' that would resemble the periodic table of chemical elements. The attempts to produce such a table continue until now (e.g. Bambach et al. 2007; Winemiller et al. 2015). The very idea is brave and welcome, indeed, as it is a high time to bring ecology and the theory of evolution closer to the exact sciences by increasingly using the method of deduction. However, I am not impressed by the progress of its realisation so far. In my scientific practice, I always prefer to start with a simple model, a general picture. I believe that the details can be filled in later. The above-mentioned colleagues, on the other hand, seem to have started the other way around. I would like to wish those who will accept this challenge in future to avoid building the 'Tower of Babel'. The starting point could be the facts regarding functional convergence of ecosystems. Their presentation should simplify the picture of the real world to such extent that the 'periodic table of niches' would fit on one page and be understandable even to a schoolchild. Of course, I have in mind the 'obligatory' trophic levels and guilds, their niches, rather than those of individual species or genera. The empty cells of the 'periodic table' could represent the adaptive zones available for immigration or evolution.

Lastly, I would like to emphasise one more aspect of evolution, namely, its dualistic nature. In Part I, I focused mainly on the positive relations between organisms, because, as it could easily be expected, they are the ones linked with the preservation and growth of biodiversity. I tried to convince my reader in Part II that evolution of the living nature is more than just appearance and unceasing development of organisms; it includes their magnificent, remarkably harmonised self-organisation into more extensive life-support systems, such as populations, communities, and ecosystems. However, we also know the other side of it – individual organisms and species enjoy a significant degree of freedom in their daily activities; along with biotic attraction and division of labour, there is also struggle for existence or biotic repulsion (Chaps. 3, 8, and 12). 'Cooperation for existence' is counterbalanced by 'struggle for existence'. By applying the principle of uniformitarianism, I attempted to demonstrate that, in the process of evolution, individuals and species have vast possibilities not only to occupy vacant niches and thus plug the gaps in the structure of communities and ecosystems but also to accumulate selfish traits and supress their competitors – in other words, to reduce diversity by keeping only the strongest 'fighters', the best-adapted species and individuals. This type of evolution could be tentatively called Darwinian evolution.

If so – and this should no longer be questionable – then both the evolution of ecosystems, so extensively discussed above, and the evolution of species, which should be understood as the accumulation of characteristics useful only at individual or species level, have been actually happening. This is a thread stretching from protobionts up to humans, allegedly 'the highest organised life form'. The dualistic nature of everyday relationships between organisms is reflected in their dualistic development. Therefore, we witness not one, as it has been traditionally considered, but two relatively independent evolutionary pathways (Lekevičius 1986, 2000, 2002). If one of them could be called Darwinian, the other one would be non-Darwinian, although that by no means implies that Darwin has not written anything about the evolution of species diversity (see Chap. 1). Needless to say, both Darwinian and non-Darwinian pathways coexist and supplement one another.

In the introductory part of the book, I wrote that Darwin encountered problems when he tried to understand whether the so-called highly organised forms are better adapted than the more primitive ones. Now, as we seem to have clarified that both evolution of species and that of ecosystems have been in action, this question is no longer a major problem. In the context of ecosystem evolution, all species in stable coexistence within the same community should be seen as equally well adapted, each to its corresponding niche consisting of biotic and abiotic elements. From the very dawn of life, species were inclined towards the division of labour, preferring complementation and cooperation to competition and elimination; therefore, any arrangement according to their status is inappropriate. However, in the context of species evolution, the 'highly organised' species indeed, as described by Darwin, are better adapted than the 'primitive' ones. Here, the species that, on their evolutionary paths, had accrued enough selfish characteristics to be able to successfully eliminate numerous other species within their guild should receive especially high ranking. Biotic attraction brings species together, all of them being equally important, as long as they sustain the material cycle and energy flow; biotic repulsion, on the contrary, sorts species into high- and low-ranking ones and – should there be no restrictions – would push biological diversity to a minimum.

Perhaps I should now remind my reader about the origins of my approach to the evolution of life. I started with an attempt aimed to clarify what are the prevailing relations between the whole and its constituent parts: species within an ecosystem and individuals within a population (Chaps. 3, 4, 5, 6, 7, 11, and 12). I concluded that there is a dynamic equilibrium ($\leftrightarrow$) in nature between biotic repulsion and biotic attraction:

Autonomy ↔ Interdependence

Struggle for existence ↔ Division of labour.

A few decades ago, by using elementary deduction I derived a regularity from this fundamental principle of ecology that might provide a suitable explanation of the appearance of biodiversity and its development. This regularity could be depicted in a similar way:

Evolutionary autonomy ↔ Evolutionary interdependence

Evolution of species ↔ Evolution of communities / ecosystems.

The next step was a thorough study of publications and empirical data. The task was not difficult, although the work lasted up until recently. Eventually, I have realised that empirical material collected by numerous authors does not seem to contradict, with some minor exceptions, my theoretical scheme (Chaps. 13, 14, 15, 16, 17, 18, 19 and 20). Of course, let the reader be the ultimate judge.

So, the specific aspect of my approach towards the development of life lies in the emphasis of the idea that evolution, and, more specifically, diversification of species, is not, and has never been, a chaotic process governed by migration, random genetic variation, climate change, or natural disasters alone. Wherever we look, be it the present times or distant geological past, a lush rainforest or a parched desert, we encounter not just a random species diversity but a well-organised diversity.

Just like any other system of views which aims to achieve the status of a theory, mine should not just explain something that has not been explained before but also offer some predictions. When we deal with the past events, such predictions are called retrodictions (Wächtershäuser 1997). According to R. Root-Bernstein (*The Scientist*, September 1997):

> Retrodictions are extrapolations into the past about things that should be (but have not yet been) found in the geological, palaeontological, or archeological record if a theory is correct. Darwin, for example, retrodicted that many intermediate forms would be found in the fossil record linking human beings and other primates and that similar intermediates would be found linking modern horses with primitive mammals. We must remember that, at the time, no such fossils were known.

Let me list some predictions and retrodictions, which originate from my theory and, if eventually confirmed by evidence, would support the theory itself (explanation of the concepts, as well as the sources and the more elaborate content of the statements, can be found above, in the main text):

1. **The existing biodiversity is the result of life's inclination towards specialisation (the division of labour)**. Species are specialised according to the material cycle and corresponding energy flows. Genotype diversity is an example of even deeper specialisation. At the same time, ecological communities and populations ensure their higher total biomass and are able to stabilise it both in terms of time

and space (by adjusting the optima) with the help of biodiversity. This suggests that the trend towards specialisation was largely determined by the variability of environmental conditions on Earth. The phenomenon of biodiversity might not be that common on the planets with more uniform and stable climatic and edaphic conditions.

2. **Every time (e.g. at the dawn of life, its colonisation of land, in the process of primary succession), nutrient cycles used to occur together with the appearance of life itself or soon afterwards.** This means that either a single species or a single genotype is not and cannot be fully autonomous, neither in terms of functioning nor development ('only an ecosystem is alive').
3. **Ecological succession and evolution are the forms of self-organisation of life, and both of them are subject to similar organisational constraints.** Ecological communities and ecosystems are specifically arranged systems with certain control and coordination. In addition, this means that each time when biodiversity increases, it tends to acquire a similar structure: similar combinations of 'functional kingdoms', trophic levels, and guilds (the phenomenon of functional convergence of ecosystems). Most species should have at least a few (in most cases – hundreds) ecological equivalents outside their usual habitats.
4. **The assemblage of ecological communities and ecosystems seems to always start 'from the bottom up' and proceed with acceleration.** Communities and ecosystems are assembled by filling in the pre-existing vacant niches and (automatically) creating new niches for the future immigrants or organisms that would evolve in situ. Self-assemblage seems to always start with producers and/or detritus and end with the occurrence of top predators and their parasites. Evolution of nutrient cycles follows a similar sequence: the pre-existing sources of energy and electrons as well as electron acceptors turn into vacant niches and sooner or later start to be used. The same happens with the metabolic waste. Thus, diversity of species is growing, and the cycles demonstrate the trend to become not only increasingly complex but also more efficient, i.e. waste free.
5. **As ecological communities and ecosystems used to become fully assembled in the course of evolution, the subsequent growth of biodiversity would have to slow down, whereas species extinction events caused by biological reasons (competition, overexploitation) would become more common.** Majority of the species that had existed in the geological past but eventually died out could not match the more modern species in terms of their adaptedness, so they were displaced. Usually this would happen when species migrated from one geographical region to another. This also implies that evolution of species has led to some progress, which (in this case) I would define as the increase of adaptedness, accrual of traits beneficial exclusively for the species itself.
6. **From the cosmological perspective, the evolution of life on Earth means life's expansion, both extensive and intensive growth of the mass of the biosphere.** Biodiversity is only a means, not the goal. Most probably, any other extraterrestrial form of life would follow the same trend of increasing its total biomass, although the means to achieve that might not necessarily be identical to ours.

7. **All species accumulate two types of traits: those useful only to themselves and those beneficial to both them and their communities.** Once a certain species manages to accrue a critical quantity of selfish traits, the species might become a supercompetitor. Numerous supercompetitors had existed in the geological past; however, only *Homo sapiens* succeeded to accrue enough of these selfish traits that it approaches the landmark of true functional and evolutionary autonomy. What this autonomy will be used for is beyond the scope of my theory. As far as I know, none of these seven points is subject to any broader discussions in the framework of mainstream biology. Moreover, most of my colleagues would hardly accept any of them as correct without reservations. However, the main problem lies not in the failure to find proper answers but in the failure to even ask questions, even those once raised by Darwin himself.

Let me summarise all that was written about the evolution of life (Chaps. 16, 17, 18 and 19) as follows:

- Since the very moment of their appearance on Earth, organisms have been growing, reproducing, and expanding their domains. The increase of total biomass appears to be the main 'goal' of life. However, on a planet with sufficiently heterogeneous environment, both in terms of space and time, this 'goal' could be achieved only by producing large numbers of specialised (adapted to the local or temporary conditions) genotypes and species. Heritable variation, 'struggle for existence', as well as directional and disruptive selection, were the tools to accomplish that.
- The first life forms on Earth were most probably producers. They already were capable of growing, reproducing, and expanding their range. However, no matter where they settled, soon they would encounter a very unfortunate circumstance: producers could not ensure their own long and stable existence all by themselves. Therefore, a strong negative feedback would kick in sooner or later, pushing towards the formation of a new functional kingdom, the decomposers. Here, again, the tools were heritable variation, 'struggle for existence', as well as directional and disruptive selection. The direction of selection would be determined by the local conditions and chemical composition of the detritus. Decomposers also had the ability to grow, reproduce, and expand their range; soon they became inseparable companions of their indirect mutualists and irreplaceable architects and participants of various nutrient cycles.
- Apparently, life was increasing its total biomass in both extensive and intensive ways. This may imply that organisms must have encountered resource deficit relatively early. Upon depletion of a certain resource, life would shift towards new, more abundant ones. Under these circumstances, functional efficiency, including that of nutrient cycles, would always become the highest priority.
- During the period between 4 and 2 billion years ago, biomass was not subject to direct exploitation as a resource. Thus, it turned into a huge adaptive zone open for potential users. Such a situation could not last forever in the environment with scarce resources. Once the primary consumers had evolved the course of events started to accelerate, diversification of multicellular organisms in the

Cambrian and Ordovician led to the formation of other trophic levels that had been missing before, and the ecological pyramids acquired an almost modern shape. The lower trophic levels produced the upper ones, because any new organism would automatically become a vacant niche for its future consumers from the moment of its appearance.

- The principles of the occurrence and self-assembly of the consumer block on land were not much different from those observed in the development of aquatic communities. Apparently, these two processes of evolutionary self-organisation of communities were shaped by the same thermodynamic and purely biological constraints.
- The engine of evolution is heritable variation and natural (Darwinian) selection; however, selection itself is subject to constraints that occur during intra- and interspecific interactions between organisms. These constraints manifested in such a way that the inconceivably broad field of possibilities sooner or later produces not random but quite specific, almost invariant combinations of genotypes, species, guilds, trophic levels, and 'functional kingdoms'. This could be the most concise and simple causal explanation of biodiversity.
- Individuals and species are endowed with a dualistic nature – they must be both adapted and co-adapted. Therefore, in addition to the selfish traits, individual organisms and species are compelled to accumulate throughout the course of their evolution also traits that are beneficial for the population and the community. The former enhance the competitive capacities of an individual or species, their adaptedness, and lead towards the elimination of the weak, while the latter result in the division of labour; they stabilise populations and communities and through them biodiversity itself.

It seems, my last task is to present my reader with an extremely simplified key to the mystery of biodiversity (see Chap. 1), which has been lingering around for over 160 years. The facts and concepts presented in this monograph allow me to think that it could be as follows:

- Biodiversity is *supported* by trophic specialisation and the diversity of environmental conditions (in other words, it is a form of the division of labour).
- Biodiversity helps to *stabilise the functions* (essential variables) of individuals, populations, and ecological communities.
- *The emergence* of biodiversity is determined by heritable variation and the advantage of specialised structures over non-specialised ones.
- Biodiversity is characterised by its ability to increase itself and to *organise itself* into relatively consistent life-support structures, which we call populations, communities, and nutrient cycles.

I call the system of views presented in this monograph ecological Darwinism because it combines Darwinism and neo-Darwinism with the modern theory of ecology.

References

Adoutte, A., Balavoine, G., Lartillot, N., Lespinet, O., Prud'homme, B., & Rosa, R. D. (2000). The new animal phylogeny: Reliability and implications. *Proceedings of the National Academy of Sciences USA, 97*, 4453–4456.

Alroy, J., Aberhan, M., Bottjer, D. J., Foote, M., Fürsich, F. T., et al. (2008). Phanerozoic trends in the global diversity of marine invertebrates. *Science, 321*, 97–100.

Bambach, R. K., Bush, A. M., & Erwin, D. H. (2007). Autoecology and the filling of ecospace: Key metazoan radiations. *Palaeontology, 50*, 1–22.

Bengtson, S. (2002). Origins and early evolution of predation. *Paleontological Society Papers, 8*, 289–317.

Blamey, N. J. F., Brand, U., Parnell, J., Spear, N., Lécuyer, C., et al. (2016). Paradigm shift in determining Neoproterozoic atmospheric oxygen. *Geology, 44*, 651–654.

Bryant, C. (1994). Ancient biochemistries and the evolution of parasites. *International Journal for Parasitology, 24*, 1089–1099.

Cavalier-Smith, T. (2002). The phagotrophic origin of eukaryotes and phylogenetic classification of protozoa. *International Journal of Systematic and Evolutionary Microbiology, 52*, 297–354.

Davidov, Y., & Jurkevitch, E. (2009). Predation between prokaryotes and the origin of eukaryotes. *BioEssays, 31*, 748–757.

De Clerck, O., Bogaert, K. A., & Leliaert, F. (2012). Diversity and evolution of algae: Primary endosymbiosis. *Advances in Botanical Research, 64*, 55–86.

Dunne, J. A., Williams, R. J., Martinez, N. D., Wood, R. A., & Erwin, D. H. (2008). Compilation and network analyses of Cambrian food webs. *PLoS Biology, 6*, 693–708.

Erwin, D. H., Laflamme, M., Tweedt, S. M., Sperling, E. A., Pisani, D., & Peterson, K. (2011). The Cambrian conundrum: Early divergence and later ecological success in the early history of animals. *Science, 334*, 1091–1097.

Gray, J. (1988). Evolution of the freshwater ecosystem: The fossil record. *Palaeogeography, Palaeoclimatology, Palaeoecology, 62*, 1–214.

Hedges, S. B., Marin, J., Suleski, M., Paymer, M., & Kumar, S. (2015). *Tree of life reveals clock-like speciation and diversification*. Molecular and Biological Evolution. Advance Access publication, March 3, 2015.

Hutchinson, G. E. (1959). Homage to Santa Rosalia or why are there so many kinds of animals? *American Naturalist, 93*, 145–159.

Javaux, E. J. (2007). The early eukaryotic fossil record. *Advances in Experimental Medicine and Biology, 607*, 1–19.

Javaux, E. J., Knoll, A. H., & Walter, M. R. (2004). TEM evidence for eukaryotic diversity in mid-Proterozoic oceans. *Geobiology, 2*, 121–132.

Knoll, A. H., Javaux, E. J., Hewitt, D., & Cohen, P. A. (2006). Eukaryotic organisms in Proterozoic oceans. *Philosophical Transactions of the Royal Society of London B, 361*, 1023–1038.

Lekevičius E. (Лекявичюс Э.). (1986). *Elements of a general adaptation theory*. Mokslas. (in Russsian).

Lekevičius, E. (2000). *Only an ecosystem is alive: A somewhat nontraditional approach to the evolution of life*. Vilnius University. (in Lithuanian).

Lekevičius, E. (2002). *The origin of ecosystems by means of natural selection*. Institute of Ecology. (Vilnius).

Lekevičius E. (Лекявичюс Э.). (2018). Biodiversity – What for and why? II. In search for a synthetic explanation. *Journal of General Biology (Moscow), 79*, 221–236. (in Russian).

LoDuca, S. T., Bykova, N., Wu, M., Xiao, S., & Zhao, Y. (2017). Seaweed morphology and ecology during the great animal diversification events of the early Paleozoic: A tale of two floras. *Geobiology, 15*, 588–616.

Martin, W. F., Garg, S., & Zimorski,V. (2015). Endosymbiotic theories for eukaryote origin. *Philosophical Transactions of the Royal Society B. 370*, 20140330. https://doi.org/10.1098/rstb.2014.0330

Miller, M. F., & Labandeira, C. C. (2002). Slow crawl across the salinity divide: Delayed colonization of freshwater ecosystems by invertebrates. *GSA Today, 12*, 4–10.
Minter, N. J., Buatois, L. A., Mángano, M. G., MacNaughton, R. B., Davies, N. S., & Gibling M. R. (2016). The Prelude to Continental Invasion. In M. G. Mángano, & L. A. Buatois (Eds.), The Trace-Fossil Record of Major Evolutionary Events. *Topics in Geobiology, 39*, 157–204.
Nooijer de, S., Holland, B. R., & Penny, D. (2009). The emergence of predators in early life: There was no garden of Eden. *PLoS One, 4*(6), e5507.
Park, L. E., & Gierlowski-Kordesch, E. H. (2007). Paleozoic lake faunas: Establishing aquatic life on land. *Palaeogeography, Palaeoclimatology, Palaeoecology, 249*, 160–179.
Pianka, E. R. (1978). *Evolutionary ecology* (2d ed.). Harper and Row.
Porter, S. (2011). The rise of predators. *Geology, 39*, 607–608.
Sanchez-Baracaldo, P., Raven, J. A., Pisani, D., & Knoll, A. H. (2017). Early photosynthetic eukaryotes inhabited low-salinity habitats. *Proceedings of the National Academy of Sciences USA, 114*, E7737–E7745.
Schopf, J. W. (1992). The oldest fossils and what they mean. In J. W. Schopf (Ed.), *Major events in the history of life* (pp. 29–63). Jones and Bartlett.
Sepkoski, J. J., Jr. (1979). A kinetic model of Phanerozoic taxonomic diversity, II: Early Phanerozoic families and multiple equilibria. *Paleobiology, 5*, 222–252.
Vannier, J. (2012). Gut contents as direct indicators for trophic relationships in the Cambrian marine ecosystem. *PLoS One, 7*, e52200.
Vermeij, G., & Lindberg, D. (2000). Delayed herbivory and the assembly of marine benthic ecosystems. *Paleobiology, 26*, 419–430.
Vidal, G. (1984). The oldest eukaryotic cells. *Scientific American, 250*(2), 32–41.
Wächtershäuser, G. (1997). The origin of life and its methodological challenge. *Journal of Theoretical Biology, 187*, 483–494.
Wheat, C. W., & Wahlberg, N. (2013). Phylogenomic insights into the Cambrian explosion, the colonization of land and the evolution of flight in Arthropoda. *Systematic Biology, 62*, 93–109.
Winemiller, K. O., Fitzgerald, D. B., Bower, L. M., & Pianka, E. R. (2015). Functional traits, convergent evolution, and periodic tables of niches. *Ecology Letters, 18*, 737–751.
Wray, G. A., Levinton, J. S., & Shapirot, L. H. (1996). Molecular evidence for deep Precambrian divergences among metazoan phyla. *Science, 274*, 568–573.
Xiao, S., & Laflamme, M. (2008). On the eve of animal radiation: Phylogeny, ecology and evolution of the Ediacara biota. *Trends in Ecology & Evolution, 24*, 31–40.
Zhang, X., Shu, D., Han, J., Zhang, Z., Liu, J., & Fu, D. (2014). Triggers for the Cambrian explosion: Hypotheses and problems. *Gondwana Research, 25*, 896–909.

Chapter 20
Supercompetitors. *Homo Sapiens* as a Consumer of Biodiversity

Our special interest should focus on the species that I would describe as supercompetitors. They are able to displace especially many other species which occupy similar niches. Photosynthetic sulphur bacteria might have been such supercompetitors at the times of their occurrence; later, cyanobacteria became one of these; even later, this role was played by the first fishes and many other organisms that had acquired radically innovative features enabling them to significantly expand their range. In all cases, these supercompetitors would inflict catastrophes potentially approaching the scale of so-called mass extinctions that might have been caused by cosmic bodies or volcanic activities. However, a thorough observer might also notice that eventually the balance of power would be restored and supercompetitors would turn into ordinary, albeit abundant, species. Along with the infliction of casualties, supercompetitors would themselves produce almost as many new species. Besides, their expansion would be followed by some kind of negative feedback, which would halt it at some point. For instance, in the case of cyanobacteria, they bumped into the free oxygen that they themselves had produced, and it also must have decimated the ranks of anaerobic decomposers, which were their indirect mutualists.

But let us explore a more familiar example that could well illustrate the course of events. One of these could be the occurrence of woody vegetation in the Devonian (see Chap. 16). It seems likely that the process started with one or several successful mutations that enabled herbs to produce woody tissue consisting almost exclusively of lignin, hemicellulose, and cellulose. The mutations increased the competitiveness of these plants, as they could develop stronger and taller stems, thus overshadowing their herbaceous relatives. The new genotypes would spread in the population to become the 'wild type', and then the population would proceed to expand at the cost of other species. As the range of the new species was growing, it started to diversify, i.e. produced new subspecies and species, thus damaging even more the plants of other communities using the same or similar niches. Thus, the first Earth's forests might have evolved. The original minor piece of genetic information contained in the genes of a single individual has eventually been amplified almost to the global scale (the role of the strengthening of genetic information in the processes of non-genetic adaptation was discussed in Chap. 11).

However, at a certain point, most likely, soon after the occurrence of the first woody plants, they must have encountered a rather serious problem (Lekevičius 2002, 2009). Namely, lignin is a very complex organic polymer without a fixed (and thus specific) primary chemical composition, because it is assembled randomly

E. Lekevičius, *Biodiversity*, https://doi.org/10.1007/978-3-031-11582-0_20

from smaller building blocks, the oligomers. Therefore, genetic variation and selection cannot easily produce matching enzymes that would effectively decompose these compounds. Without proper enzymes, a tree trunk may rest in the soil for millennia without decaying. Some early herbaceous plants might have produced lignin but in much smaller quantities. In such a situation, a lot of waste, in the form of undisintegrated wood, had accumulated in the material cycle. Tree trunks would pile up, covering up most the land surface, and the major part of nutrients could not be recycled. A strong negative feedback had formed and threatened to eliminate both the mutants and their ecosystems. This situation might have led to the formation of large coal layers in the end of the Devonian and throughout the Carboniferous Period (Robinson 1990; Hibbett et al. 2016).

According to some data (Floudas et al. 2012; although see Nelsen et al. 2016), this vacant niche provided by lignin was occupied with a great delay in the end of the Carboniferous, some 295 million years ago, with the occurrence of fungi specialised to decompose lignin, most likely from among Basidiomycota. Thus fungi, as indirect mutualists, helped woody plants and might have even saved them from extinction.

The emphasis should fall on the ultimate result of this hypothetical course of events: as forests became widespread, the total biomass of terrestrial ecosystems has increased by at least several tens of times in the span of a few or a dozen million years. Evolution of the woody plants had transformed into evolution of ecosystems, and selfishness was harnessed to achieve common 'goals'. This story, therefore, pretty well illustrates the interaction between the two evolutionary pathways mentioned earlier. Even more importantly, it demonstrates that ecosystems or communities do not necessarily have to be subject to selection as separate units in order to evolve.

Nevertheless, in the geological past, neither a single supercompetitor species nor even a guild consisting exclusively of supercompetitors had ever succeeded to become functionally independent, i.e. autonomous to the extent that they could start and finish the nutrient cycle, thus closing the loop. Both cyanobacteria and the first woody plants, as we know, were not even able to resist the appearance and expansion of their more recent competitors (algae and seed plants, correspondingly) without losses. Therefore, one may suspect the existence of a certain cycle that would start with the rise of a supercompetitor and end with its transformation into an ordinary species (Fig. 20.1).

Homo sapiens is also a supercompetitor, in fact, the most competitive of all the species that have ever existed. It might be reasonable to assume that the longer is the evolutionary history of a species, the more selfish characteristics it might have accrued. The human is one of the most recent species; therefore its history is one of the longest.

As it could be deduced from Darwin's theory, humans should be characterised by not only the lowest death rate (*d*) among terrestrial organisms but also the lowest birth rate (*b*) per capita. Thus, they enjoy the highest reproductive efficiency ($E_r = r/b$, where *r* is population growth rate per individual; see Lekevičius (1986, 2002, or 2013) for more details on this indicator). Nowadays, the percentage of

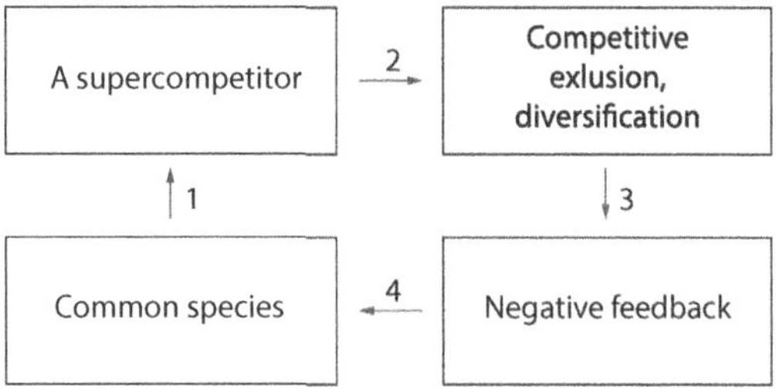

Fig. 20.1 The cycle of appearance, expansion of a supercompetitor, and its transformation into a common species

newborns who successfully reach reproductive age is approximately 95–98% in the developed countries and 50–80% in the least developed countries. This is a rough estimate of the efficiency of human reproduction in percentages. As a comparison, the E_r of large birds and mammals is no bigger than 20–30%, that of reptiles 1% at most, of amphibians 0.01%, and of fishes usually not exceeding 0.0001%. I should add that this evolutionary trend was described also by some other researchers, including Simpson (1949, pp. 257, 285):

> There is a sequence from mere emission of eggs and sperm into the environment, (…) through internal fertilisation, to various degrees of care for the fertilized eggs until the young are born, and finally to postnatal care for the young (…) His (human – E. L.) reproductive efficiency is the highest in the animal kingdom, with prenatal care at least as high as in any other animal and postnatal care decidedly higher.

The modern human species is also special because it is the ultimate supercompetitor, and there will be no more similar precedents in the biosphere. The fate of other supercompetitors might not necessarily apply to humans. Apparently, our species has both taken over from our evolutionary ancestry and accrued by themselves so many qualities enhancing our competitive advantage, that the delicate balance between biotic attraction and repulsion has been disrupted for long (even in the context of geological time scale) and, possibly, irreversibly. Throughout the relatively short history of our species, humans seem to have aimed all their efforts towards the elimination of their main competitors. We have become not just ordinary supercompetitors but real consumers of biodiversity. All the previous supercompetitors had behaved in a similar way, but the scope of their impact used to be much smaller. Besides, we have exterminated the predators that would pose a threat to us, and now we put significant effort into achieving the same with our parasites. Finally, we rearrange landscapes and habitats at will, exterminating thousands of species – not only those that we consider pests but, even more massively, those that would never cause us any harm.

No other supercompetitor has ever set such goals, neither were they able to. Humanity expected that these desperate attempts to transform the ecological niche inherited from its evolutionary ancestry would be successful, especially with the expansion of scientific and technical revolution, with the invention of pesticides, vaccines, and antibiotics. Even nowadays, many of us remain convinced that humans are on the winning course.

On top of that, we managed to colonise practically all climatic zones of our planet in an unusually short time span and by involving only microevolutionary mechanisms (races). *Homo sapiens* did not even have time to split into separate species or subspecies adapted to different climatic conditions. The fate of other supercompetitors was different. The progenitor population of woody plants that emerged in the Devonian might have taken several million years to spread across the continents, and during this slow expansion, it must have produced a high number of derivative species of trees and shrubs. A similar diversification was part of the reptilian expansion in the Permian.

Nature has responded to the growth of human population by increasing parasitic pressure, and predators would have played their own role, had we not exterminated them. This was supposed to happen as our species was moving ever closer to becoming a vacant niche. Genetic defects are accumulating at a high rate in human populations (Rühli and Henneberg 2013). This could also be expected after stabilizing selection had been relaxed to a minimum (Kolcov 1924; Lekevičius 1986, 2002). It should be noted, too, that humanity has not yet succeeded to create waste-free production, a closed material cycle similar to those existing in nature. There is no doubt that this problem will become more and more pressing, even if we do not perceive its full scope right now. This could be anticipated bearing in mind that, just as the previous supercompetitors, humans have only been enjoying the seemingly unlimited freedom to reproduce and to do whatever they want so far.

It is quite possible that in the distant future, all the negative feedback, no matter how numerous, will be successfully eliminated with the help of science and technology, and species evolution will completely overwhelm evolution of ecosystems. Ecosystems and the biosphere, as we know it, will cease to exist and will be replaced by a structure which some call the anthroposphere – others the noosphere or the technosphere. Maybe some future evolutionary biologists will conclude that species diversity and ecosystems which had persisted for millions of years were only subsidiary measures, a provisional system, which was necessary to create conditions for the appearance of *Homo sapiens*, who then sprouted like a seed lurking in the womb of Nature and waiting for the right moment. Maybe they will proudly declare: our history is a success story, which finally embodies the ultimate strive of all species that have ever existed – to become functionally independent, to conquer nature and subjugate other species for their own interests… or to eliminate them completely. Maybe, indeed, *Homo sapiens* will become the first and the last species which can suffice with adaptation alone and does not need coadaptation?

Can such a scenario be real? I think so, at least we seem to be heading in this direction. But will humanity be able to survive without biodiversity? As a biologist, I can tell by experience that yes, in principle, it could. Maybe I and you, my reader, – we would not manage, but people, their needs and interests change from generation to generation. They are able to withstand many things and to adapt, if necessary. Our planet receives enough energy from space; another rich source is stored deep in its entrails. The Earth contains enough chemical elements and compounds that we need, and they should not disappear anywhere in the future. So, we would most likely be able to feed ourselves, even without plants and animals, relying on future technologies instead. Maybe we could even increase the efficiency of this production – after all, plants utilise only 1–2% of the total light energy that reaches the leaf surface. Waste handling might be a bit more of a challenge, but even here solutions could be found. So far, we simply have shown little effort to address the difficulties in this field. Futurologists try to assure us that the parasite problem will also be resolved in the future by simply exterminating them 'once and forever'. Let us assume this will happen. What else could be exterminated? Oh yes, there are also decomposers, 'the smallest among brothers of ours'. On the other hand, why destroy them, especially bearing in mind that the task would be far from easy? We could keep them, might say a supporter of the 'anthropo-noo-techno-sphere', let them handle the problem of waste and recycle various materials.

This is a sketch drawn by some futurologists who have trust in human nature and wisdom. I personally am not so sure about both human nature and wisdom. Therefore, I would not recommend to frivolously reject yet another possible scenario, that of Apocalypse. The chance that humanity will self-destroy or be destroyed by some cosmic catastrophe is far from impossible. That could give a start to a new wave of self-organisation, which eventually, millennia from now, from the dust and the remnants of the biosphere would produce new intelligent beings – hopefully, wiser than me and you…

In any case, the human has taken over from nature the control of evolution, and this probably means that voluntarist actions of just one species, or even a single leader, might have a stronger impact on the biosphere than the laws of nature that have been in effect until recently. Forecasting the future of our planet by simply extrapolating the previous course of development might be irresponsible. The ancient and also the modern history of humanity has convincingly proven that we can hardly trust 'His Majesty Intellect' (*Noos* in Greek). Sounds incredible, but the intelligent human, *Homo sapiens*, can transform into an aggressive and dangerous creature in a flash. So many times before, the 'naked ape', as described by the famous British ethologist Morris (1967), would turn into just an ape armed with a razor; therefore, trying to predict its behaviour is a tricky task. A certain paradox can be noticed in the history of nature and the human race: The spontaneous forces of the soulless nature have been acting in a rather coordinated way to create an unparalleled diversity of living creatures and enchanting harmony, whereas the purposeful actions of *Noos*, it seems, may be good only for turning this harmony into ashes.

Not long ago, another book by Wilson (2012), whom I quoted a few times, appeared, and this one is dedicated to the human phenomenon. While reading it, I was delighted to notice how far we have moved forward since the times of Darwin's *The Descent of Man*. Thanks to Wilson, today we have a much clearer picture of who we are, as *Homo sapiens*, where we differ from the rest of nature and where we are similar, where we have come from and where we are heading. Regretfully, the overall picture that he presented leaves no reason for optimism – his conclusions sound almost like a severe verdict (Wilson 2012, pp. 7, 13, 62, and 244):

> Humanity today is like a waking dreamer, caught between the fantasies of sleep and the chaos of the real world. The mind seeks but cannot find the precise place and hour. We have created a Star Wars civilization, with Stone Age emotions, medieval institutions, and

> god-like technology (…) We are an evolutionary chimera, living on intelligence steered by the demands of animal instinct. This is the reason we are mindlessly dismantling the biosphere and, with it, our own prospects for permanent existence. Humanity is a magnificent but fragile achievement (…). Our bloody nature, it can now be argued in the context of modern biology, is ingrained because group-versus-group (tribe-versus-tribe, state-versus-state – E. L) was a principal driving force that made us what we are (…) Our instincts remain unprepared for civilisation.

War is 'humanity's *hereditary* (emphasized by E. L.) curse', according to this author. What does he mean? Most likely he meant that our inner nature and emotional profile have remained almost unchanged through these millennia. Possibly, because we acquire knowledge and skills in the course of our lives, and we can transform them swiftly, while our nature and passions, unfortunately, are passed as an unchangeable heritage from generation to generation.

In my opinion, E.O. Wilson has reached such intellectual heights in many fields of biology that hardly any modern biologist could compete with him. Therefore, I tend to trust him as an unparalleled authority. Nevertheless, he has acquired numerous opponents in the recent years. On the other hand, that alone does not mean much – a prevailing opinion is seldom the best one in the situations that require exceptional insights. By the way, some time earlier, Morris (1967) also received a similar reaction. His version of the portrait of *Homo sapiens* was slightly different from that of Wilson, but it appears equally pessimistic:

> Optimism is expressed by some who feel that since we have evolved a high level of intelligence and a strong inventive urge, we shall be able to twist any situation to our advantage; that we are so flexible that we can re-mould our way of life to fit any of the new demands made by our rapidly rising species-status; that when the time comes, we shall manage to cope with the over-crowding, the stress, the loss of our privacy and independence of action; that we shall remodel our behaviour patterns and live like giant ants; that we shall control our aggressive and territorial feelings, our sexual impulses and our parental tendencies; that if we have to become battery chicken-apes, we can do it; that our intelligence can dominate all our basic biological urges. I submit that this is rubbish. Our raw animal nature will never permit it. (Morris 1967, pp. 210–211)

Many other outstanding personalities, including the already mentioned L. von Bertalanffy and A. Koestler, present grim forecasts regarding the future of humankind. We might find even more pessimists among the world-famous writers and philosophers, i.e. the most sensitive and deep-thinking personalities, who register the slightest shifts in the moods or priorities of a modern human. Many of them warn that technological progress does not make people better; rather on the contrary, it grants a lot of free time to individuals with limited creativity without offering any adequate means to cope with their boredom. Thus, the 'progress' backfires and causes such a chaos in many souls which is unprecedented and leads towards unpredictable consequences.

Under the influence of all these pessimistic authors, I had a really grim vision of my own: what if the cunning and daring alpha males, hungry for power and blood, have always been more successful from prehistorical times, and they have produced statistically more progeny than their compliant, empathic, and peaceful tribesmen? And this pattern might have persisted through thousands of generations! This thought alone makes me shiver – are we really doomed? Can it be that even today we continue to spread hatred and madness across our Earth, instead of cherishing what we proudly call culture and associate with the future of our children? I am not sure one can stay cool, with the head full of these thoughts, while listening to the optimists' speeches about *Noos* and the bright future of humankind...

The only relief is in the assumption that neither the pessimism of one camp nor the optimism of the other one is based on knowledge; that they reflect something else:

> what Paul says about Peter tells us more about Paul than about Peter. (Baruch Spinoza)

A frequently asked question is whether the appearance of intelligent beings similar to humans is programmed in the very essence of the living nature, or the occurrence of humans was merely a coincidence, an extremely rare one in the astrophysical context. If the latter is true, the possibility of finding other beings similar to us on other celestial bodies would be highly unlikely. Let me sum up my personal position on this matter. If we speak about the presence of life on other planets, then, in my opinion, the answer should be a rather firm yes: life does exist beyond our Earth. As far as other intelligent beings are concerned, contrary to Conway Morris (2003), I do not have an equally firm opinion. Although, with a generous margin of doubt, I would again support those who are positive on this. Even less likely seems the chance that these beings are similar to us in terms of their appearance, intellect, values, or technologies. At least my 'theory' has nothing to say about that.

However, I am much more confident with another statement: If all goes well to our species, *Homo sapiens* will disappear within a few millennia by way of pseudo-extinction, as natural selection (or, rather, the lack thereof) will transform our species into a new one.

References

Conway, M. S. (2003). *Life's solution: Inevitable humans in a lonely universe*. Cambridge University Press.

Floudas, D., Binder, M., Riley, R., Barry, K., Blanchette, R. A., & Henrissat, B. (2012). The Paleozoic origin of enzymatic lignin decomposition reconstructed from 31 fungal genomes. *Science, 336*, 1715–1719.

Kolcov N. K. (Кольцов Н. К.). (1924). The influence of culture on selection in humanity. *Russian Journal of Eugenics, 2*(1), 3–19. (in Russian).

Lekevičius, E. (2002). *The origin of ecosystems by means of natural selection*. Institute of Ecology.

Lekevičius, E. (2013). *Ecology: From individuals to the biosphere*. Vilnius University. (in Lithuanian).

Lekevičius E. (Лекявичюс Э.). (1986). *Elements of a general adaptation theory*. Mokslas. (in Russian).

Lekevičius E. (Лекявичюс Э.). (2009). On some analogies between ecosystem evolution and development of economy: From A. Smith and C. Darwin to the latest ideas. In L. E. Grinin, A. V. Markov, & A. V. Korotaev (Eds.), *Evolution: Cosmic, biological, and social* (pp. 226–260). Librokom. (in Russian).

Morris, D. (1967). *The naked ape*. Jonathan Cape Publishing.

Nelsen, M. P., DiMichele, W. A., Peters, S. E., & Boyce, C. K. (2016). Delayed fungal evolution did not cause the Paleozoic peak in coal production. *Proceedings of the National Academy of Sciences USA, 113*, 2442–2447.

Rühli, F. J., & Henneberg, M. (2013). New perspectives on evolutionary medicine: The relevance of microevolution for human health and disease. *BMC Medicine, 11*, 115.

Simpson, G. G. (1949). *The meaning of evolution*. Oxford University Press.

Wilson, E. O. (2012). *The social conquest of earth*. Liveright Publ. Corp.

Chapter 21
Has Darwin's Theory Really Become Obsolete?

I shall try to briefly define my position regarding neo-Darwinism. By this term, I mean all of its modifications, including: 'modern synthesis' (1930s–1950s), 'extended synthesis' (from 1980s) and 'expanded synthesis' (from 2004?) – see Maynard Smith and Szathmáry (1995), Kutschera and Niklas (2004), Ridley (2004), Futuyma (2009, 2015), Pigliucci (2009). Neo-Darwinism is usually presented as a synthesis of Darwinism, traditional and population genetics, and also palaeontology (see Laland et al.(2015) for a more comprehensive list and additional information). Evolutionary ecology, another discipline which exists from around 1970 till now (Pianka 1978, 2000), despite rather pretentious title, has no unique character, as I see it; therefore it could also be mentioned along with the other versions of neo-Darwinism listed above.

Like Laland et al. (2015), I also think that the theory of evolution is far from complete. First, as I wrote in Chap. 2, my proposed methodology of synthesis is significantly different from that currently prevailing in evolutionary biology. Naturally, the derived conclusions and the vision of the evolution of life are also different. There are at least several points where our opinions diverge:

- On the position regarding selection factors and characteristics favoured by selection
- On the evolution of ecosystems as one of the evolutionary pathways
- On the role of competition in evolution
- On the understanding of trends and progress in macroevolution

Let us begin with the first two as they are closely connected. I do not support those (e.g. Williams 1966; Dawkins 1976), who state that a gene rather than a phenotype is a unit of selection. I see it just the other way around: genetic variation offers, whereas the phenotype and its environment perform the selection; genes are not directly involved in either cooperation or struggle for a place in the gene pool; they are information carriers and, as such, cannot be attributed to the kinetic sphere of an organism. Therefore, they can be considered neither altruistic nor selfish. I might be wrong, but I think that no other idea has been more detrimental to evolutionary biology than this one. Even more so, because it remains surprisingly persistent even nowadays, especially among biologists who have only a shallow understanding of the problems addressed by evolutionary biology. These biologists are convinced that a gene or a genotype can be – and indeed is – more active than the phenotype.

E. Lekevičius, *Biodiversity*, https://doi.org/10.1007/978-3-031-11582-0_21

Let us recall the story of this revision of Darwinism. Back in the first half of the previous century, a proposal was submitted to replace the concept of struggle for existence with the concept of the 'struggle' of alleles (genes) for a place in the gene pool. This led to the appearance of population genetics and the perception of evolution as a shift in gene frequencies within populations. Later, some biologists came up with the idea that the laws of microevolution, i.e. variations of gene frequencies within population, could be easily extrapolated to the laws of macroevolution, perhaps just by adding the concept of reproductive isolation mechanisms and their development. This was nothing else but logical reduction; however, as we can see now, an inexcusably crude one, because it left no place for ecology and organismal biology in the theory of evolution. Once the ecological part was dropped (some considered this a well justified step), the conviction that species evolution and macroevolution have no clear direction or trends and are governed by random processes has grown stronger.

Many biologists have been warning that such revision of Darwinism was too crude and unnecessary. For instance, already 50 years ago, Waddington (1968), one of the most outstanding theoreticians, wrote this:

> The most precisely formulated statements of evolutionary theory are enshrined in the mathematical theories of Haldane, Fisher and Wright (...) The basic abstractions or simplifications of reality, however, which they used to enable them to cast the problem into a particular set of differential equations still underlie and limit all the more recent developments. Very few physicists and not many more biologists realize just how limiting these simplifications are.

In spite of that, the gene-centred reasoning turned out to be surprisingly persistent. With the appearance of new disciplines of genomics and metagenomics, or community genomics, it seems to have gained a second wind. These new fields of study are linked with high expectations, including the possibility to finally resolve the mystery of biodiversity. Nonetheless, I think that these dreams are nothing but illusion. I fully support Strohman (1997), a rather well-known expert of molecular biology, who states:

> ... I think that the Watson-Crick era, which began as a narrowly defined and proper theory and paradigm of the gene, has mistakenly evolved into a theory and a paradigm of life, that is, into a revived and thoroughly molecular form of genetic determinism. (...) today, in most universities, especially in the United States, if you are a biologist and are not cloning genes, you are mostly unfunded (...) A consensus is created in which 'normal science' provides further insights into the molecular-genetic mechanisms of life. Alternative paradigms such as systemic or epigenetic biology languish. A profound shift occurs in our perception of the world (from organisms to gene machines) in which we learn more and more about mechanisms and less and less about life.

Wilson (2017), whom I mentioned and quoted quite often, made a similar assessment:

> Complete genomes make possible quick scans of entire faunas and floras. They also permit the reconstruction of the evolutionary history by which related species have multiplied. Yet in the broader perspective of biodiversity, these studies are the equivalent of aerial surveillance; what is more needed are boots on the ground. The ongoing neglect of biodiversity research impedes the progress of conservation of life at all levels in all taxonomic groups. It also diminishes the capacity to meet one of the greatest challenges to the biological

> sciences, rising just over the horizon: the origin, evolution, and equilibration of ecosystems. The problems presented by ecosystem analyses are equivalent in complexity to those presented by the human brain. They can be solved by nothing less than a Linnaean renaissance, in which each one of the millions of Earth's species still surviving is discovered and its role in the biosphere increasingly well documented.

In these words, I see the wisdom based on deep historical knowledge typical of this author. So, the gene-centred view of the world might be simple, but it is completely inadequate. I believe, Darwin was right by saying that an individual organism should be considered a *key* unit of selection. A population, or even community, could also become such a unit, if the degree of 'interconnection' or integration between its constituent individuals and species were high enough. Therefore, there are reasons to believe, as some researchers do (Reeve and Hölldobler 2007; Wilson 2008; Nowak et al. 2010), that specific groups of individuals, such as local societies of termites, ants, wasps, or bees, are also involved in differential survival. Indeed, they have acquired many characteristics of a superorganism. The researchers do not insist that individual selection has lost its importance in such societies. Instead, they believe that these societies are subject to what they call a multilevel selection mechanism. See, e.g., Flinn et al. (2005) for the discussion on how this selection might have facilitated the cognitive evolution of human species. Quite logically, Wilson (2012) included humans into this list of social organisms (pp. 52, 289):

> Multilevel selection consists of the interaction between forces of selection that target traits of individual members and other forces of selection that target traits of the group as a whole (…) groups compete with groups, favoring cooperative social traits among members of the same group. At the lower level, members of the same group compete with one another in a manner that leads to self-serving behavior. The opposition between the two levels of natural selection has resulted in a chimeric genotype in each person. It renders each of us part saint and part sinner.

This approach to the mechanism of evolution demonstrates the main idea, which I have faithfully pursued for over 30 years (Lekevičius 1986, 2011, 2019): if individuals (and species) are not functionally autonomous, they cannot be autonomous also in their evolution; then, besides the ordinary individual selection, there must be another type of selection that changes characteristics of groups (and ecological communities). In other words, individuals (and species) must be subject not only to adaptation but also to coadaptation. Genotype suggests, phenotype selects; phenotype suggests, population selects; population suggests, ecological community selects (Lekevičius 1986). To achieve that, there is no need for the selection of groups and communities, as units. I chose to call this form of selection 'cascade selection', instead of 'multilevel selection', because I think that this name reflects its essence much better.

Other examples of 'non-Darwinian' selection (Gould 1982; Wilson 1997; Gould and Lloyd 1999) are far less successful, in my opinion. In the worst case, one could consider two sexual partners as a unit of selection, since they are also 'linked together', but I would rather not, in order to avoid confusion: sexual partners undergo individual variations from generation to generation, although they are also compelled to coevolve, thus restricting each other's individual evolution. The same could be said about plants and their pollinators: they guide each other's heritable variation so that coadaptation is not lost. In both cases, the entire system evolves as a united front, although selection affects individuals. A similar logic is applicable to the evolution of ecosystems as units, although the acting forces are not limited to mutualism in this case.

To conclude, one may assert that the problem of the units of selection has been so messed up by researchers themselves (e.g. Futuyma 2009) and that it could keep a few more generations of evolutionary biologists preoccupied. Which is a pity... Should anybody ask me, I would advise this: dear colleagues, let us cease analysing what is being selected, and let us focus on what determines selection instead. This may be a somewhat trickier but much more essential question. Let me refer those readers who are interested to learn more about my approach to natural selection, its nature, and forms to my earlier publications (e.g. Lekevičius 1986, 2003, 2011, 2019).

Meanwhile, let us move to the second point. Darwin was right when he stated that natural selection discards the weaker 'strugglers for existence' produced by heritable variation. This should lead to the conclusion that these 'strugglers for existence' become more and more advanced with each generation, i.e. they become better adapted to resist competitors, predators, parasites, and climatic conditions as well as more efficient in finding food. Since, in Darwin's understanding, not only conspecifics but also coexisting species struggle for existence, it would be logical to expect that biodiversity would shrink from generation to generation. Perhaps, it would not have even emerged, in the first place. However, we observe just the opposite trend in real life: diversity has been increasing. This could mean that either the concept of struggle for existence is wrong (which would contradict the obvious facts) or an offsetting force of biotic attraction has also been acting in nature besides struggle.

Should that be the case, then selection would discard not only the weaker *competitors* but *also* inappropriate *partners*, and it especially promotes those whose contribution is essential for the survival and reproduction of separate individuals or the entire species. Therefore, characteristics of individuals, but also of populations and communities, change from generation to generation. In other words, most populations and practically all 'natural economies' (i.e. communities or ecosystems) are organised structures, which means that they impose rather strict requirements on evolutionary variation of their constituent parts, which are species. The destiny of organisms is to be simultaneously adapted and coadapted, and it seems that so far nature has not been offering any alternative solution. I do not expect many evolutionary biologists of the modern generation to support this idea, because in their understanding populations and ecosystems are not organised systems. They interpret cooperation and mutualism as the outcome of long evolutionary process, rather than selection factors that have existed from the very appearance of life on Earth.

This leads to another divergence of our positions. According to the prevailing opinion, the expressions 'evolution of ecosystems (communities, biosphere)' are inadequate since these structures cannot evolve. Allegedly, only populations and species are evolving, while the changes in community characteristics are nothing but a side effect of this evolution. This position was inspired by a widespread, albeit misleading, conviction (e.g. Williams 1966; Gould 1982; Wilson 1997; Gould and Lloyd 1999) that only the structures that could be identified as units of selection can evolve; naturally, communities and ecosystems cannot be reasonably considered such units (and there is no reason to disagree with this statement). I, nonetheless, insist that communities and ecosystems do evolve, just like species. This happens by way of biotic interactions within a population and community, which act as a powerful selection factor. It is during the interactions between organisms that the regular Darwinian selection is given a specific direction. And it is this interaction that steers evolution so as to produce organised systems instead of chaotic assortments of species that would turn into the arena of struggle for existence. The evolution of life on Earth is a long-lasting process of self-organisation of populations, communities, and the biosphere. A similar process was the self-organisation of economic systems (Leigh and Vermeij 2002; Lekevičius 2009): competition involves private individuals and their groups linked through cooperative relations; meanwhile, the country's economy is evolving. As Vermeij (2004, p. 1) put it very precisely, the evolution of life is determined by 'competition-driven demand and resource-driven supply'.

Let us move on to the third point. The divergent role of competition seems to cause doubts neither to me nor to my real or imaginary opponents. Everybody, apparently, agrees that competition acts as a centrifugal force pushing species towards segregation and division of labour. However, in fully assembled communities, a peaceful division of niches is not always possible. Suspecting this or similar situation, Darwin wrote that struggle for existence can be, and often might have actually been, the reason of species extinction; the stronger and better adapted species displace the weaker ones. This is one of the key postulates of Darwinism, and it can easily be deduced – after all, not only conspecifics are involved in this struggle. Space and resources are not limitless, and extinction provides space and resources for a new, more advanced species. In Darwin's opinion, only such elimination of less adapted forms can ensure the realisation of evolutionary progress, i.e. development from more primitive towards highly organised, and thus more advanced, forms.

As I have already mentioned elsewhere (Chap. 17), an alternative opinion is becoming predominant, namely, that competition has been playing a minor role in the extinction of species and that most taxons died out as the result of some cosmic or geological disasters; only after another extinction event would clear some niches, other taxons would take them over (the passive replacement hypothesis). I can only add the following in response to this opinion. First, it can hardly be considered an extension or clarification of Darwinism. Rather, it resembles a return to the catastrophism theory of G. Cuvier (1769–1832), which Darwin considered unacceptable. I may be wrong, but my impression is that the emphasis on the impact of abiotic events is usually placed by palaeontologists (maybe quite unintentionally), while quite many biologists still stick to the Darwinist positions. This is understood since the opinions of researchers are largely shaped by their main field of education. Moreover, both camps lack firm evidence, which is difficult to obtain.

However, this is not the key point. If competitive displacement never happened, would any progressive development, as it was understood by Darwin, be possible altogether? An economist could ask a similar question: could a country's economy develop successfully without competition between individuals and associations (companies, firms, or factories) and the risk of bankruptcy? I think that most of us would answer negatively to both these questions. Does that mean conclusions offered by the catastrophism theory are wrong? I think so.

Strangely enough, some evolutionary biologists seem to have chosen another path, which appears rather unexpected to me – they deny the existence of any lasting large-scale trends and evolutionary progress. In other words, they try to prove that the organisms which had evolved later and survived till our times do not have any advantages in comparison with the extinct species (which must have included quite many ecological equivalents) – see, for instance, the overview by Ruse (1996). Let us admit – this is indeed a radical solution, especially bearing in mind that, besides Darwin himself, the idea of progressive development was embraced by practically all his successors, including such outstanding evolutionary biologists of the twentieth century as Huxley (1942, 1958) and Grant (1977). Although, admittedly, there were some sceptics as well (e.g. Williams 1966; Lewontin 1968; Simpson 1973).

I think that the currently prevailing position regarding this issue of key importance is well reflected by Futuyma (2009, pp. 603, 605), one of the greatest authorities in evolutionary biology:

> Darwin (...) imagined that if long-extinct species were revived and had to compete with today's species, they would lose the competition badly. We might find evidence of the evolution of competitive superiority if the fossil record provided many examples of competitive *displacement* of early by later taxa – but as we have seen (in Chap. 7), this pattern is much less common than *replacement* by later taxa well after the earlier ones became extinct (...) We cannot meaningfully compare the level of adaptedness of a shark and of a falcon (...) Progress in the sense of betterment implies a value judgment, and here we must guard against the parochial, anthropocentric view that human features are better than those of other species (...) The great majority of animal lineages (...) show no evolutionary trend toward greater 'intelligence' (...) it is hard to document even objective trends...

The author's reference to Chap. 7 mostly implies the works by Rosenzweig and McCord (1991) and by Benton (1996).

As far as I understand, the reasoning of the opponents of evolutionary progress is as follows: if the species coexisting in well-established communities are equivalent in terms of their ability to struggle for existence, then there is a high probability that all species from the distant past, including the extinct ones, could also match them. Such a statement, however, rejects the very core of Darwinism.

No wonder that to many of us evolution has become an unpredictable process governed only by random coincidences. The idea that selection does not follow any stable long-term trend was expressed already by Mayr (1968), one of the founders of neo-Darwinism. Gould (1994), another famous classic, presented even clearer summary of this position:

> Natural selection is a principal of local adaptation, not of general advance or progress. The history of life is not necessarily progressive; it is certainly not predictable. The earth's creatures have evolved through a series of contingent and fortuitous events.

Such an attitude is unacceptable to me. In my understanding, by rejecting the option which allows for relatively frequent cases of competitive displacement, we lose any reason to talk about evolutionary *development* of species altogether. I have never had any doubts (Lekevičius 1986) whether evolution has any direction or trends, or not. The expressions 'non-directional evolution' or 'evolution driven by random coincidences' clearly are obvious oxymorons. In that case, what does natural selection do with non-directional heritable variation? 'Non-directional evolution' basically implies that for millions of years life has been meandering around one and the same point in the evolutionary landscape without actually getting anywhere. Now, this indeed would be an inexplicable, even absurd, statement. Of course, random events always affect the final result. But that hardly means that they *govern* such processes as the evolution of life. When I use the expression 'random events', I mean not only genetic variation or gene drift but also geological or cosmic factors of various scale, which, according to many quoted researchers, had essentially predetermined the evolution of life. In my opinion, a certain determining impact of such events cannot be denied, especially that of cosmic disasters, motions of the Earth's crust, volcanic activity, and even glaciations. But they do not govern evolution, and thus cannot set any relatively stable trajectory for the development of life.

By the way, this position shared by some serious and well-known neo-Darwinists seems to suit very well all sorts of mystics, especially the supporters of Intelligent Design. Because if evolution does not have a fixed course and is governed by Her Majesty Randomness, then how could it have produced such a perfect creation, which the human undoubtedly is, without the aid of transcendental powers; and not just the human, but the entire grand harmony of Nature, above all? The promoters of Intelligent Design have come to a logical conclusion that this would have not been possible without Somebody who has a 'super-brilliant' model of the creation of life that is way beyond the limits of human understanding. And it has been implemented with extreme diligence.

So, I can state with regret that the currently prevailing doctrine has not only failed to assimilate the understanding of ecosystem evolution, it also rejects certain key postulates of Darwinism. A curious and rather intriguing detail: this revision of the fundamentals of Darwinism, carried out by modern evolutionary biologists, is not advertised as such. While reading articles or modern textbooks (e.g. Ridley 2004; Futuyma 2009), few readers would realise that the author or authors deny the essential postulates of Darwinism. Such disguise, possibly even *conspiration du silence*, could be understood as the authors' reluctance to tarnish the name of Darwin, the most prominent figure in our field of study. Most likely, this is the case. However, the written words cannot be denied, unless they were meant to conceal the actual ideas. I think such 'discretion' is unnecessary in serious science. *Amicus Plato, sed magis amica veritas.*

On the other hand, it is much more difficult to tell which large-scale trends or patterns did actually exist, as well as which of them were essential and which played a supplementary role. Many evolutionary biologists of the twentieth century have been thinking about that (e.g. Huxley 1942, 1958; Grant 1977; Simpson 1949; see also the overview by Rosslenbroich 2006). The following were included among the proposed criteria of evolutionary progress: complexity of the anatomy (increasing trend), degree of differentiation of tissues and organs (increasing trend), competitiveness (increasing trend), resistance to extinction (increasing trend), body size (increasing trend), metabolic rate (increasing trend), degree of cephalisation and ability to reason (increasing trend), reproductive efficiency (increasing trend), functional efficiency (increasing trend), fecundity and mortality rate (decreasing trend), and a few more. In short, many experts have had no doubts regarding the presence or absence of wide-ranging trends. However, the diversity of these criteria in published sources is so high that it most probably indicates the lack of consensus on this matter, rather than the abundance of 'goals' typical of life itself. Besides, some of these criteria are not universal, they might apply to the animal kingdom, but not to plants or microorganisms.

We may find a possible solution in the general adaptation theory (GAT; see Chap. 11). It states that the efforts of all adaptation mechanisms, including the evolutionary ones, are aimed in the same direction, namely, towards the increase of biomass by any means and in any circumstances, be it the biomass of a species or the total biomass of a community. This trend has been ensured not only by heritable variation and selection but also by ever-increasing biodiversity and functional efficiency (see Chap. 7 and Fig. 11.5). Life had no other way to expand its realm but through specialisation and diversification of species and genotypes. Thus, diversification was merely a tool rather than the goal. I believe that this particular 'goal' – namely, extensive and intensive growth – is and will be pursued by any life, not just its terrestrial forms. Unless different thermodynamics and different physical constants are at play in other parts of the Universe. Again, few of my colleagues would support the opinion that adaptedness can be measured in terms of biomass – it may sound unexpected, crude or even peculiar. On the other hand, I am prepared to change my opinion as soon as somebody presents convincing arguments to deny it. So far, however, nobody has managed to convince me otherwise.

The position that increase of the total biomass of the biosphere reflects perhaps the main 'goal' of life on Earth seems to agree pretty well with the perception of evolution by experts in thermodynamics (e.g. Wicken 1980). By the way, already Simpson (1949, pp. 243, 245), one of the most outstanding palaeontologists of the last century, has expressed a similar idea:

> In considering the record of life, we sought progressive changes that involve life as a whole, and not only the evolution of particular groups within the total process. Only one was found: a tendency for life to expand, to fill in all the available spaces in the livable environments, including those created by the process of that expansion itself. This is one possible sort of progress. Accepting it as such, it is the only one that the evidence warrants considering general in the course of evolution (…) We have seen, too, how at any given time certain groups tended to be much more varied and abundant than others, in other words, to dominate the life scene, and that there has been a succession in these dominant groups. It is this succession that provides a criterion, though plainly not the only one, of evolutionary progress (…) Aside from the broad tendency for the expansion of life, which is also inconstant, there is no sense in which it can be said that evolution is progress.

These words could also belong to Darwin, as they reflect pretty well his position that, in the struggle for existence, the stronger, and therefore better adapted, competitors are those that displace the weaker ones and occupy their place.

To conclude, I would like to emphasise the thought that Darwin's ideas are far from obsolete and it would be too early to 'discard' them. According to Reznick and Ricklefs (2009), to many contemporary biologists, *The Origin of Species* is nothing more but 'a beautiful fossil', a work of merely historical value. The authors of that article convincingly demonstrate that some of Darwin's ideas of far from secondary importance are either misunderstood or forgotten, although nobody has proven them wrong (see also Boero 2015). This is exactly what I tried to achieve throughout the pages of this monograph. Even more, I attempted to bring my reader back to *The Origin of Species* and invited to study this epochal work with great diligence, without skipping a single paragraph. I did not mean to say that it holds answers to all the questions we encounter. There was no way Darwin could resolve the mystery of biodiversity, even though he tried. We, the followers of his ideas, shall have to complete this task by ourselves. My attempt to achieve that may not be the most successful one; however, being the first of its kind, it might serve as an icebreaker for other, more creative evolutionary biologists. In any case, I consider my approach closer to Darwinism than to neo-Darwinism and 'the Evolutionary Synthesis'.

How come that we, biologists, have drifted so far away from the fundamentals of Darwinism during those 160 years? One possible explanation could be that, again and again, we have been failing to overcome the curse of complexity and, especially, to resolve the mystery of biodiversity; therefore we opted for the easiest although not the most rational way: we have divided the world into billions of allegedly independent components and made them accessible for analytical, particularly, mathematical methods. However, by ignoring the *relations*, we made the living world enormously complex and even less understandable. This dealt the ultimate blow to the hope that an overarching biological theory could ever be created. We ought to have acted the other way around. *The world is an integrated whole; therefore, it is simple.*

References

Benton, M. J. (1996). On the nonprevalence of competitive replacement in the evolution of tetrapods. In D. Jablonski et al. (Eds.), *Evolutionary Paleobiology* (pp. 185–210). University of Chicago Press.

Boero, F. (2015). From Darwin's origin of species toward a theory of natural history. *F1000Prime Reports 2015, 7*, 49. https://doi.org/10.12703/P7-49

Dawkins, R. (1976). *The selfish gene*. Oxford University Press.

Flinn, M. V., Geary, D. C., & Ward, C. V. (2005). Ecological dominance, social competition, and coalitionary arms races: Why humans evolved extraordinary intelligence. *Evolution and Human Behavior, 26*, 10–46.

Futuyma, D. J. (2009). *Evolution* (2nd ed.). Sinauer Associates, Inc.

Futuyma, D. J. (2015). Can modern evolutionary theory explain macroevolution. In E. Serrelli & N. Gontier (Eds.), *Macroevolution* (pp. 29–85). Springer.

Gould, S. J. (1982). Darwinism and the expansion of evolutionary theory. *Science, 216*, 380–387.
Gould, S. J. (1994). The evolution of life on the earth. *Scientific American, 271*, 63–69.
Gould, S. J., & Lloyd, E. A. (1999). Individuality and adaptation across levels of selection: How shall we name and generalise the unit of Darwinism? *Proceedings of the National Academy of Sciences USA, 96*, 11904–11909.
Grant, V. (1977). *Organismic evolution*. W. H. Freeman and company.
Huxley, J. S. (1942). *Evolution: The modern synthesis*. George Allen and Unwin.
Huxley, J. S. (1958). Evolutionary processes and taxonomy with special reference to grades. In O. Hedberg (Ed.), *Systematics of today*. Uppsala Universitets Årsskrift.
Kutschera, U., & Niklas, K. J. (2004). The modern theory of biological evolution: An expanded synthesis. *Naturwissenschaften, 91*, 255–276.
Laland, K. N., Uller, T., Feldman, M. W., Sterelny, K., Müller, G. B., Moczek, A., Jablonka, E., & Odling-Smee, J. (2015). The extended evolutionary synthesis: Its structure, assumptions and predictions. *Proceedings of the Royl Society. Ser. B, 282*, 20151019.
Leigh, E. G., Jr., & Vermeij, G. J. (2002). Does natural selection organize ecosystems for the maintenance of high productivity and diversity? *Philosophical Transactions of the Royal Society of London B, 357*, 709–718.
Lekevičius E. (Лекявичюс Э.). (1986). *Elements of a general adaptation theory*. Mokslas. (in Russian).
Lekevičius E. (Лекявичюс Э.). (2003). Ecosystem evolution: The major stages and possible mechanisms. *Journal of General Biology (Moscow), 64*(5), 371–388. (in Russian).
Lekevičius E. (Лекявичюс Э.). (2009). On some analogies between ecosystem evolution and development of economy: From A. Smith and C. Darwin to the latest ideas. In L. E. Grinin, A. V. Markov, & A. V. Korotaev (Eds.), *Evolution: Cosmic, biological, and social* (pp. 226–260). Librokom. (in Russian).
Lekevičius, E. (2011). Ecological Darwinism or preliminary answers to some crucial though seldom asked questions. In L. E. Grinin, A. V. Korotaev, R. L. Carneiro, & F. Spier (Eds.), *Evolution: Cosmic, biological, and social* (pp. 101–121). Uchitel Publishing House.
Lekevičius, E. (2019). Ecological Darwinism or preliminary answers to some crucial though seldom asked questions (Lekevičius 2011, reprinted). In L. Grinin & A. Korotaev (Eds.), *History and mathematics: Big history aspects* (pp. 100–121). Uchitel Publishing House.
Lewontin, R. C. (1968). Evolution. In D. L. Sills (Ed.), *International encyclopedia of the social sciences* (Vol. 5, pp. 202–210). The MacMillan Company and The Free Press.
Maynard, S. J., & Szathmαry, E. (1995). *The major transitions in evolution*. W. H. Freeman.
Mayr, E. (1968). Cause and effect in biology. In C. H. Waddington (Ed.), *Towards a theoretical biology* (Vol. 1, pp. 42–54). Prolegomena. An IUBS Symposium. Edinburgh: Edinburgh University Press.
Nowak, M. A., Tarnita, C. E., & Wilson, E. O. (2010). The evolution of eusociality. *Nature, 466*, 1057–1062.
Pianka, E. R. (1978). *Evolutionary ecology* (2nd ed.). Harper and Row.
Pianka, E. R. (2000). *Evolutionary ecology* (6th ed.). Benjamin-Cummings.
Pigliucci, M. (2009). An extended synthesis for evolutionary biology. *Annals of the New York Academy of Sciences, 1168*, 218–228.
Reeve, H. K., & Hölldobler, B. (2007). The emergence of a superorganism through intergroup competition. *Proceedings of the National Academy of Sciences USA, 104*, 9736–9740.
Reznick, D. N., & Ricklefs, R. E. (2009). Darwin's bridge between microevolution and macroevolution. *Nature, 457*, 837–842.
Ridley, M. (2004). *Evolution* (3th ed.). Blackwell Publ.
Rosenzweig, M. L., & McCord, R. D. (1991). Incumbent replacement: Evidence for long-term evolutionary progress. *Paleobiology, 17*, 202–213.
Rosslenbroich, B. (2006). The notion of progress in evolutionary biology – The unresolved problem and an empirical suggestion. *Biology and Philosophy, 21*, 41–70.
Simpson, G. G. (1949). *The meaning of evolution*. Oxford University Press.

Simpson, G. G. (1973). The concept of progress in organic evolution. *Social Research, 41*, 28–51.
Strohman, R. C. (1997). The coming Kuhnian revolution in biology. *Nature Biotechnology*. 15 March 1997.
Vermeij, G. J. (2004). *Nature: An economic history*. Princeton University Press.
Waddington, C. H. (1968). Towards a theoretical biology. *Nature, 218*, 525–527.
Wicken, J. S. (1980). A thermodynamic theory of evolution. *Journal of Theoretical Biology, 87*, 9–23.
Williams, G. C. (1966). *Adaptation and natural selection*. Princeton University Press.
Wilson, D. S. (1997). Biological communities as functionally organized units. *Ecology, 78*, 2018–2024.
Wilson, E. O. (2008). One giant leap: How insects achieved altruism and colonial life. *Bioscience, 58*, 17–25.
Wilson, E. O. (2012). *The social conquest of earth*. Liveright Publ. Corp.
Wilson, E. O. (2017). Biodiversity research requires more boots on the ground. *Nature Ecology and Evolution, 1*, 1590–1591.

Afterword

Life science is in deep crisis nowadays, just like physics was over a century ago. Enchanted by the analytical method, we, biologists, still struggle to lift the curse of complexity and to find the appropriate forms of logical simplification. Back in 1911, physicists took a wise step – they organised the first Solvay Conference. The entire constellation of greatest physicists was invited, 23 persons in total. The participants discussed the ways to pull their discipline out of the crisis that it was in. By joining efforts, they have found a solution, even if not right away.

I think it is high time for us, biologists, to follow their example. We have gone a long way to split the objective wholeness into subjective separate components, but this is hardly an achievement we should be proud of – the world has become even more complicated. I think that, as the years go by, we are increasingly losing our hope to ever understand *how Nature operates in its natural environment*. In a few more generations, this opportunity may be lost forever because *this* Nature with *these* interactions will simply be gone. A few more years, and the last of those who were able to see the living nature and biology as integral systems will also be gone for good.

E. Lekevičius, *Biodiversity*, https://doi.org/10.1007/978-3-031-11582-0

Index

E. Lekevičius, *Biodiversity*, https://doi.org/10.1007/978-3-031-11582-0

The manufacturer's authorised representative in the EU is Springer Nature Customer Service Centre GmbH, Europaplatz 3, 69115 Heidelberg, Germany. If you have any concerns regarding our products, please contact ProductSafety@springernature.com

Printed and bound by CPI Group (UK) Ltd, Croydon, CR0 4YY
28/09/2026
02224091-0009